# Food Innovation and Engineering 2018

AA009258

Topical Conference at the 2018 AIChE Annual Meeting

Pittsburgh, Pennsylvania, USA
28 October - 2 November 2018

ISBN: 978-1-5108-7644-6

**Printed from e-media with permission by:**

Curran Associates, Inc.
57 Morehouse Lane
Red Hook, NY 12571



Printed by Curran Associates, Inc. (2019)



AIChE
120 Wall Street, FL 23
New York, NY 10005-4020

Phone: (800) 242-4363
Fax: (203) 775-5177

www.aiche.org

**Additional copies of this publication are available from:**

Curran Associates, Inc.
57 Morehouse Lane
Red Hook, NY 12571 USA
Phone: 845-758-0400
Fax: 845-758-2633
Email: curran@proceedings.com
Web: www.proceedings.com

# TABLE OF CONTENTS

**Some abstracts in this book refer to images that were not made available by the publisher**

# 539584 Pittsburgh: Urban Agriculture and the Food-Energy-Water Nexus


**Abstract:**

Cities rely on external supply chains for food, energy and water. Urban areas are growing and the dependence on reliable supply chains for these resources grows. The interdependence between the infrastructure and connections of providing these resources has become increasingly significant. Urban agriculture provides an opportunity to address Nexus challenges. Urban agriculture benefits include increasing food resilience and improving stormwater management. Ancillary benefits can also include the potential to reduce the amounts of materials sent to landfills (through composting), the reduction urban heat-island effect, and reducing wastewater treatment loads â€“ all of which come with energy reduction benefits.

Pittsburgh lost a substantial portion of its population during the transition away from its industrial roots as the â€˜Steel City.â€™ A legacy of the population decrease includes empty and abandoned lots (e.g. over 8,000), aging infrastructure, and areas of decline that have created food deserts and challenges associated with the interdependence of food, energy and water. Pittsburgh has implemented multiple initiatives to address this challenge that will realize the benefits from urban agriculture (e.g. a 23 acre urban farm; community driven revitalizations that include energy efficient housing, stormwater infrastructure, rainwater management, and agriculture spaces).

This analysis examines urban agriculture strategic options, the potential benefits, and the challenges. Options include agriculture options (e.g. geoponic, hydroponic, aeroponic systems), water management options (e.g. water reuse), waste management (e.g. biochar as soil enhancer and carbon storage), and energy option (e.g. distributed generation).

# 538328 Quantifying Virtual Phosphorus Flows in Interstate Food Trade: Implications for Environmental Sustainability

**Abstract:**

Phosphorus(P) is one of the essential nutrients for agriculture. It is primarily supplied to crops as inorganic phosphorous fertilizer or though manure application. On farm use of phosphorus efficiency is very low, ranging between 10-15%. Rest of the applied P gets bonded in the soil which may subsequently become available for crop uptake. Current agronomic P recommendations for farmers exclude the accumulated P due to unknown timescales of release. Sustainable management of P is an important topic due to the limited availability of inorganic P sources and its downstream potential to cause eutrophication. Additionally, production of phosphorus fertilizers is highly energy intensive with limited potential of waste stream recovery due to techno-economic constraints. Therefore, an accurate understanding of regional P application and efficiency, combined with interregional food trade is necessary to understand the complexity of P cycle. We combine data on current phosphorus application practices with recommended and existing soil P levels to assess sustainability of P management on farm. Additionally, we use P application data with food trade to quantify phosphorus embodied in the inter-regional trade. This work will enable to identify trends and hotspots in US agriculture P application practices for management of food-water nexus.

We develop a food trade and associated virtual phosphorus model for the U.S. by evaluating over 60 crop items covering commodity groups of cereal grains, fruits, nuts, and vegetables. Regional phosphorus mass balances are constructed using datasets from US Department of Agricultureâ€™s Natural Resource Conservation Service, International Plant Nutrition Institute, and Census of Agriculture. The interstate food trade network is formulated from Freight Analysis Framework and translated to virtual phosphorus network by using crop yields and phosphate fertilizer application rates from Census of Agriculture. Preliminary results indicate that phosphorus fertilizer application differ considerably for crop ranging from 2-10 lb P per ton crop for grains to 10-68 lb P per ton of crop for nuts. To estimate whether calculated P application rates are excessive or necessary for specific soil type, we compare it with recommended P levels in each state. In 2012, 28 states had to recommended levels, including states with highly eutrophic waterbodies such as Ohio, Florida, California, Texas, and Maine. With addition of total P fertilizer across crops in the given year, the soil phosphorus for these states increases by a fraction of 2-4 compared to recommended levels. From food trade perspective, we find that 13 billion US tons of food and 5.3 million US tons of embodied P were transferred across the U.S. Specifically, grains dominate food and total embodied P transfers within the U.S., indicating the importance of considering scale of production. The implications of the study findings will be discussed in the context of balancing on-farm conservation policies for simultaneously enriching soil with nutrients and avoiding excess run-offs.

# 533778 Towards a Two-Level Superstructure Optimization Framework for Land Use Based on Food-Energy-Water Nexus

**Abstract:**

Land use optimization is the result of competitions between different land types [1]. The main reasons of these competitions are the quantitative constraints of land scales and corresponding Food-Energy-Water (FEW) resources under multiple conflicting objectives [2, 3]. Despite FEW resources play critical important roles for sustaining and improving human life, increasing production demands and sustainable concerns intensify these competitions due to their limitation. How to coordinate these competitions is a complex task for reasonable land allocation based on limited FEW resources while promoting efficiencies and sustainability [4, 5]. A nexus thinking by considering the FEW flow and interactions in multi-objective land use systems has become apparently necessary for efficient use of FEW resources and trade-off decisions [6, 7]. However, challenges arise in representing quantitative FEW-Nexus that supports land-use decisions while encountering conflict objectives, limited data, and multiple scales [8, 9].

To address these challenges, we develop a multi-objective optimization framework by considering total profit, food production, resources use, and environmental penalty as multiple objectives for land use systems. The proposed framework suggests a two-level superstructure optimization method: (1) In the first level, models of all the production units are developed by data-driven modeling and global optimization methods based on limited realistic data. FEW flow among them are quantified and interlinked to construct interval models, which can be represented as interval superstructures. The small-scale MINLP problems can be solved efficiently due to the limited combinations of land units in the interval subsystems; (2) In the second level, multiple interval models with optimal land and FEW allocations are used to construct the extended systematic network and represented as a large-scale superstructure, which can be solved as a MILP problem. A series of FEW indices are provided for decision-makers to analysis nexus in the system, carry out quantitative assessment based on different objectives, and achieve trade-off solutions. The framework is illustrated by a case study on the crop-livestock system, which shows that it can provide multiple land allocation solutions and predicts corresponding yields for land units within the systems under flexible scales. Computational results from interval models indicate that we achieve optimal solutions in the interval subsystems, and valuable production models for yield prediction with high confidence. The performance of these models can be improved by increasing feedback data [10]. The systematic network provides an efficient method for extending the land allocation solutions to multiple land scales. For multiple objectives, the proposed FEW index can be applied to select strategies for optimal land allocation that maximizing food productivity and minimizing water and energy consumptions.

**Reference**

[1] Memmah, M. M., Lescourret, F., Yao, X., & Lavigne, C. (2015). Metaheuristics for agricultural land use optimization. A review. Agronomy for sustainable development, 35(3), 975-998.

[2] Bergstrom, J. C., Goetz, S. J., & Shortle, J. S. (2013). Land use problems and conflicts: Causes, consequences and solutions. Routledge.

[3] Keairns, D. L., Darton, R. C., & Irabien, A. (2016). The energy-water-food nexus. Annual review of chemical and biomolecular engineering, 7, 239-262.

[4] Miralles-Wilhelm, F. (2016). Development and application of integrative modeling tools in support of food-energy-water nexus planningâ€"a research agenda. Journal of Environmental Studies and Sciences, 6(1), 3-10.

[5] Simpson, G., & Berchner, M. (2017). Water-energy nexus-Measuring integration: towards a water-energy-food nexus index. Water Wheel, 16(1), 22-23.

[6] Garcia, D. J., & You, F. (2016). The water-energy-food nexus and process systems engineering: a new focus. Computers & Chemical Engineering, 91, 49-67.

[7] Ringler, C., Bhaduri, A., & Lawford, R. (2013). The nexus across water, energy, land and food (WELF): potential for improved resource use efficiency?. Current Opinion in Environmental Sustainability, 5(6), 617-624.

[8] McCarl, B. A., Yang, Y., Schwabe, K., Engel, B. A., Mondal, A. H., Ringler, C., & Pistikopoulos, E. N. (2017). Model Use in WEF Nexus Analysis: a Review of Issues. Current Sustainable/Renewable Energy Reports, 4(3), 144-152.

[9] McCarl, B. A., Yang, Y., Srinivasan, R., Pistikopoulos, E. N., & Mohtar, R. H. (2017). Data for WEF Nexus Analysis: a Review of Issues. Current Sustainable/Renewable Energy Reports, 4(3), 137-143.

[10] Nie, Y., Avraamidou, S., Li, J., Xiao, X., & Pistikopoulos, E. N. (2018). Land use modeling and optimization based on food-energy-water nexus: a case study on crop-livestock systems. 13th International Symposium on Process Systems Engineering; Elsevier; Accepted.

# 537263 Land Availability, Utilization, and Intensification for a Solar Powered Economy

**Abstract:**

Among all forms of renewable energy, solar energy is the most promising one with sustainable availability and abundance to meet human beingâ€™s energy need. However, solar energy is a diffuse resource and harvesting it requires vast tracts of land. While most studies estimated that land availability is not a major challenge for the feasibility of a total solar economy, a few recent detailed analyses have raised the problem in land availability in some densely populated countries, using actual solar farm production data and considering other factors such as storage requirement and intermittency of solar availability. However, according to their estimation, countries with relatively low population density, such as the United States, have sufficient land for solar farms to meet all energy demand. In this study, we perform a detailed land requirement analysis in each state of the US for a solar powered future and propose land intensification strategies to resolve possible land use challenges.

Land requirement analysis in this study takes into account several issues that are usually ignored. First, efficiencies of major energy conversion steps from current primary energy to end use are different from those when solar energy is deployed. Second, available land is limited to only part of the urban area (rooftops) and miscellaneous land. Farmland, forest and other reserved land such as national and state parks cannot be utilized for solar plant construction. In addition, we prefer local photons for local use for consideration of minimizing transmission loss and energy security.

Our land requirement analysis shows that for many states there is insufficient currently available land and that other land areas will need to be used, which is contrary to previous report. We propose that farmland is the best choice for land use intensification as it minimizes additional environmental impact to existing habitat and propose several photovoltaic systems for this purpose. We then estimate the power output of these systems and estimate the amount of farmland for each state that is needed to meet energy needs. By land intensification, the energy demand of each state in the US can be locally satisfied.

# 529950 Microbial Community Profile Versus Water Quality in Urban Watersheds

**Abstract:**

Microbial community profiles can serve as an indicator for unbran water quality. The aim of this study was to gain insight into bacterial diversity and infer water quality based on selected operational taxonomic units (OTUs). Here we report bacterial communities of 12 watersheds sampled at different frequencies from year 2013 to 2016, were characterized by sequencing the V4 to V5 regions of 16S rRNA genes using 454 pyrosequencing and paired-end sequencing platforms. An nMDS plot of all samples revealed that bacterial communities of the same watershed tended to cluster together and a few watersheds shared similar communities. Effects of temporal variation were more pronounced on bacterial communities within watersheds ($P < 0.01$; 2-Factor PERMANOVA) than spatial variation. Cyanobacteria of *Microcystis* and *Cylindrospermopsis* dominated certain watersheds that tended to have algal blooms. Pairwise correlation of gut and sewage-associated OTUs revealed a strong positive correlation among *Enterobacter*, *Klebsiella* and *Escherichia* and *Clostridium sensu stricto 8*, suggesting that presence of these taxa as possible indicators of impairment. A comparison of the abundances of 13 potential antibiotic resistant bacteria, revealed four watersheds with samples that correlated with higher abundances of potential antibiotic resistant bacteria compared to the other watersheds. Thus microbial community determined by high-throughput sequencing in conjunction with qPCR can be used to indicate quality of water in urban watersheds.

# 528536 Mechanisms Whereby Microbes Promote Intermediate Soil Moisture Content

**Abstract:**

The function of microbial communities is inextricably with the soil microstructure. At the pore scale, water content regulates hydraulic connectivity and thereby modulates microbesâ€™ access to aqueous and gaseous substrates. Furthermore, soil bacteria near plant roots produce extracellular polymeric substances (EPS), a polymeric substance that helps modulate soil moisture and promote plant growth. EPS regulates soil moisture (i) by serving as hydrogel, swelling during wet conditions and remaining hydrated during dry conditions, (ii) by promoting soil particle aggregates via capillary forces, (iii) by altering soil surface properties through creating water repellent surfaces. EPS-mediated moisture retention behavior in soil is complicated but can be understood by systematically controlling the microscale environment. Here, we show the effect of physical microstructure on EPS-mediated moisture retention. We employ physically complex emulated soil micromodels as well as micromodels with uniform macropore or micropore geometries. EPS was collected from stationary-phase *Sinorhizobium meliloti* cultures and suspended at different concentrations in growth media salts or in artificial groundwater. Experimental results showed that the 0.25× EPS solution dried eight times slower than deionized water, and the 1× EPS solution dried 16 times slower than deionized water in identical soil micromodels. Also, the residual saturation of 0.25× and 1× EPS solutions were 6 ± 1% and 35 ± 21%, compared with a residual saturation of 0 for deionized water. We observed no effect on rate or extent of moisture loss for the micromodels with a uniform macropore geometry. Studies with uniform micropore geometry confirmed that higher EPS concentrations were associated with a reduce rate and increased variance of moisture loss at the pore scale. In this study, we provide a systematic method to evaluate the performance of EPS solutions in modulating soil moisture, and show how EPS concentration decreases the rate of water loss and also amplifies the variability of moisture content at the pore scale. Longer term, we anticipate that the impact of this work will be to aid in the development of sustainable agriculture biotechnology for increased food production using less water.

# 528379 Nitrogen Efficient Fertilizer Materials

**Abstract:**

Nitrogen cycle management is a critical challenge in modern engineering. $NH_3$ synthesis from $N_2$ consumes several percent of total world energy while only a fraction of the synthesized fertilizer nitrogen, such as urea, is absorbed by the plants. Majority of nitrogen initially confined in urea is decomposed and released in the environment as gaseous $NH_3$ or soil and water mobile nitrate, $NO_3^-$, resulting in low sustainability nitrogen fixation process. Furthermore, for a complete plant nutrient all of the nutrients (N,P,K,Ca,Mg,S) need to be supplied in one solid granulated or liquid mixture. Aqueous liquid fertilizers are expensive to transport since their dissolved nutrient content is governed by the solubility limits.

Molecular crystals of urea with inorganic acids ($H_3PO_4$ and $H_2SO_4$) as well as solid salts ($CaSO_4$, $MgSO_4$, $Ca(H_2PO_4)$ and $Mg(H_2PO_4)$) present a convenient form of high nutrient content fertilizers that can be inexpensively transported to the remote locations and applied directly or via fertigation. Importantly, these urea inorganic acid adducts have been shown to result in a much more efficient nitrogen uptake by plants combined with significantly lower nitrogen losses. The work will be presented that will explore design of these fertilizers that contribute to the enhanced sustainability of the chemical processing. In particular, mechanochemical synthesis methods that avoid use of liquid solvents that would otherwise necessitate evaporation or result in high aqueous solution transportation costs will be presented. Since dissolution of the solid urea adducts for fertigation requires in depth knowledge of the corresponding phase diagrams, these data will also be presented. Finally, molecular structure effects of the urea adducts on the improved nitrogen cycle management will be discussed as well as the synthesis of industrial and anthropogenic waste derived nitrogen efficient molecular fertilizers.

# 539589 An ESC-Based Test for High Throughput Screening of Embryotoxicity of Drugs and Chemicals

**Abstract:**

Toxicity on embryonic development is a severe adverse reaction of drugs and chemical compounds. About 10% birth defects are caused by mutagenic, embryotoxic, and teratogenic chemicals and drugs taken by pregnant women. Therefore, embryotoxic effects of drugs need to be evaluated before clinical trials. Currently, animal-based test, embryonic stem cells (ESCs) based test (EST), rat whole-embryo culture test, zebrafish embryonic development test and the micro-mass test are applied for assessing drug embryotoxicity. EST using two cell lines is the most promising tool for *in vitro* embryotoxicity test because no animal is sacrificed. However, large scale application of EST is limited because of the long culture period and need for morphological endpoint analysis which requires experience. To predict embryotoxicity rapidly and precisely, high throughput EST could be achieved by using reporter genes and monitoring their signals from ESCs cultured in microbioreactors. Although luciferase has been used as the reporter gene in previous studies, it requires luciferin as the enzyme substrate and yet can only provide endpoint data. In this study, we have developed an embryotoxicity test based on drug responses of murine embryonic stem cells (mESC) cultured in three-dimensional polymer scaffolds in microbioreactors using enhanced green fluorescent protein (EGFP) as the reporter, which provides a non-invasive and real-time monitoring method for high throughput drug screening. Expressions of EGFP are driven by selected promoters associated with human genes involved in the regulation of proliferation and/or differentiation. Stable mES cell lines expressing EGFP under the control of these promoters were developed and used to evaluate various chemicals with known cytotoxicity and embryotoxicity validated by the European Center for the Validation of Alternative Methods (ECVAM). The results showed that our EST method could accurately predict toxicity and embryotoxicity of tested drugs, demonstrating its application for high throughput embryotoxicity screening.

# 537610 Functional Role of Bacteria Involved in Cocoa Fermentation Processes According to a Metabolic Prediction Using 16S rRNA Reads


**Abstract:**

The fermentation of cocoa beans is a natural and spontaneous process where a succession of bacteria and yeasts transforms the plant material. The coloration, acidity, size and the content of flavor precursor metabolites, such as free amino acids, organic acids, and volatile compounds, define the quality of the cocoa bean. Previous studies have linked the presence and dynamics of microbial groups with the quality of the fermented cocoa bean. The microbial groups present in the cocoa fermentation are Lactic Acid Bacteria (LAB) and Acid Acetic Bacteria (AAB); the bacteria succession is related to the physicochemical characteristics of the ecosystem (temperature, pH, and availability of nutrients). The presence of bacteria and yeast has been studied with culture-dependent methods such as DGGE, although few studies use Next Generation Sequencing technologies to characterize bacterial communities in the fermentation process. Furthermore, a revision of the metabolic pathways during the process of cocoa bean transformation has not yet been carried out.

One approach for the detection of bacteria into system is the generation of 16S rRNA sequences. With this method, it is possible to know the bacterial groups and their relative abundance during cocoa bean fermentation. Likewise, from these data, it is possible to make a prediction of gene families into the system. In this work, we propose to recognize the changes in the abundance of genes associated with metabolic pathways in different stages o f the fermentation processes. For this, a cacao farm was selected in three agroecological regions of Colombia (Montaña Santandereana, Zona Marginal Baja Cafetera, and Valles Interandinos Secos) in two seasons of 2016. Cocoa samples were collected in the upper and middle zone of a fermenter box, during fermentation process every 12 hours until the end of the process. After sequencing with Illumina MiSeq of 16S rRNA reads, the detection of taxonomic groups was performed and PICRUST was used to predict the number and abundance of gene families for each fermentation time point. Subsequently, an analysis of the metabolic pathways to which the gene families are associated was carried out using KEGG database. This methodology is an approximation to an enrichment analysis of metabolic pathways.

*Lactobacillus cacaonum*, *Lactobacillus plantarum*, and *Acetobacter pasteurianus* were detected in this study and selected as starter cultures to obtain high quality cocoa. Other groups were *Erwinia tasmaniensis*, *Komagataeibacter oboediens*, and *Klebsiella quasipneumoniae*. We obtained 3.331 gene families of which only 19,6% (654 KEGG orthology) are associated with a metabolic pathway, in greater numbers to *Global and overview maps*, *Metabolism of cofactors and vitamins* and *Amino acid metabolism*. On the other hand, 2.454 reactions were found in the KEGG database. The gene families in the metabolic pathways were associated to Glycolysis, Amino acid biosynthesis and, the flavonoid pathway. No difference were detected between the gene familes found into the fermentative processes of the agro-ecological regions. There are some pathways such as *Valine, leucine* and *isoleucine degradation*, which presented a greater proportion of gene families from the 48th hour until the end of fermentation process in the three agroecological regions. This type of analysis allows to detect and quantify genes related to Carbohydrates and Energy metabolism of bacteria functional groups during a fermentation process. Metabolic pathways as Flavonoids biosynthesis are not abundantly found, this show us that this method do not have high precision to study other metabolic pathways of interest related to flavour of cocoa fermented bean.

# 534093 Quantifying the Effect of Minimal Processing on the Kinetics and Antimicrobial Resistance of *Listeria* in Structured Food Model Systems Enriched with Natural Microflora

**Abstract:**

## Introduction and Aim:

Nowadays, there is an increasing consumer demand for minimally processed foods which retain their nutritional content, natural colouring, taste and texture.[1] To achieve such food quality, minimal food processing techniques for microbial decontamination, such as treatment with natural antimicrobial compounds, are of great interest.[2 ] One such compound is nisin, which is produced by lactic acid bacteria and is approved by the Food and Drug Administration (FDA) for use in the food industry as a preservative.[3] Additionally, minimal processing treatments can be combined with non-thermal processing, such as ultrasounds, which can have a synergistic effect and act as a hurdle for the inactivation of bacteria, thereby increasing the processing efficiency.[4] However, the efficacy of such antimicrobial compounds, especially in combination with non-thermal processing methods for food decontamination, is still unclear.[5] Furthermore, these techniques are less harsh than classical (thermal) processing methods (sterilisation, pasteurisation), so there is the potential for conditions to present a mild, sublethal stress which could induce an adaptive response in bacteria, allowing post-treatment survival. This is of concern for pathogenic bacteria such as *Listeria*, which is often associated with ready-to-eat foods and with high mortality rates. An increase in antimicrobial resistance (AMR) in *Listeria* species has also been reported in recent years, which raises the concern that effective treatment of listeriosis infections may become compromised in the future.[6]

Most available studies on the efficacy of minimal processing with natural antimicrobials and/or with non-thermal technologies on inactivating food-related pathogens (including *Listeria*) are conducted in liquid broth systems, even though most food products are solid or solid(like) e.g. soft cheeses, meats. Cells grown as colonies in a solid system experience a significantly different environment in comparison to planktonic growth in a liquid system. More specifically, diffusional limitations of oxygen and nutrients and accumulation of (acidic) metabolic products around the colony can cause a self-induced (acid) stress that may affect the microbial kinetics and the microbial environmental response.[7,8,9] Furthermore, microorganisms could display a different level of AMR development due to the general environmental stress adaptation and cross protection.[10,11,12] Bacteria found in foods are also likely to grow in co-culture with other bacteria which can result in stress due to competition and/or stress due to 'foreign' extracellular metabolic products. For example, lactic acid bacteria such as *L. lactis* are found in similar foods to *Listeria*, and are natural producers of nisin, which, as mentioned above, is an antimicrobial compound of interest for use in microbial decontamination techniques.[13] Therefore, in order to achieve a fundamental understanding of the stress response and AMR development of *Listeria* in food systems in response to novel food processing (i.e. nisin combined with ultrasounds), in the presence of other bacteria, it is important to conduct kinetic experiments in model systems that mimic as accurately as possible the microstructure and microflora of solid/solid(like) foods.

*The present work is a systematic comparative study of the microbial dynamics of Listeria, as influenced by the presence of natural microflora such as nisin-producing lactic acid bacteria, on solid(like)/structured systems of various complexities at a range of optimal and suboptimal growth temperatures.*

The model systems are created using Xanthan gum (XG) and/or Whey protein (WPI): both are used in the food industry and are stable at a wide range of temperatures, so they are good surrogates to mimic a range of real food products.

**Materials and Methods:**

Food model systems were prepared using Tryptic Soy Broth supplemented with 0.6% w/v Yeast Extract (TSBYE), with (1) no added gelling agent (planktonic growth), (2) 5% w/v XG (monophasic system), (3) 10% w/v WPI (monophasic system), or (4) a combination of 5% XG and 10% WPI, thus producing a biphasic complex food model system.

Growth kinetics at 10°C, 30°C and 37°C were quantitatively monitored for *L. innocua* ATCC 33090 in monoculture, and in co-culture with *L. lactis* NZ9700 (nisin-producing) or *L. lactis* NZ9800 (non-nisin-producing) grown as planktonic cells in TSBYE, or as surface colonies on the monophasic/biphasic food model systems described. The Baranyi and Roberts mathematical model[14] was fitted to the experimental data, and the values of the model parameters for all systems under study were analysed and compared.

The effect of ultrasound on the inactivation of *L. innocua* stationary phase colonies, either in mono- or co-culture with both *L. lactis* strains, was investigated using an ultrasonic bath at a frequency of 20kHz and power of 20W, for a treatment time of 5 minutes. Inactivation kinetics were monitored, and the inactivation model of Geeraerd et al. was fitted to the experimental data.[15]

The spatial distribution and size of stationary phase colonies on the surface of each system, both in mono- and co-culture, was investigated for all conditions under study using a commercial scanning laser confocal microscope (CLSM). Phase separation in the XG/WPI biphasic system was visualised using Rhodamine B, which binds to the protein phase only allowing for definition between the two phases. *L. innocua* colonies were visualised using DAPI, while a Green Fluorescent Protein variant of *L. lactis* was used to visualise these colonies and thus differentiate between species.

**Results and Discussion:**

The growth kinetics, i.e. lag phase, exponential growth rate and/or stationary phase, of *Listeria* were affected by the structure of the solid(like) systems and by the presence of *L. lactis* (both strains), and were different to the liquid. Differences were also observed in the growth kinetics of the nisin- and non-nisin-producing *L. lactis* strains, both in mono- and co-culture. These differences may affect the production rate and/or quantity of nisin, and thus also affect its efficacy as a microbial decontamination method. Additionally, the action of nisin on *Listeria* was different depending on the food model system and the incubation temperature, and the combination of nisin with ultrasonic treatment resulted in differences in the inactivation kinetics, i.e., inactivation rate, and shoulder period.

The use of CLSM techniques showed differences in microbial spatial organisation. More specifically, in monoculture, *L. innocua* and both *L. lactis* strains demonstrate preferential growth on the protein phase of the biphasic gelled system. By contrast, in co-culture both *L. lactis* strains grow on the Xanthan gum phase, while *L. innocua* spreads through both phases. *This is the first reported observation of such changing preferential growth in a biphasic, co-culture system.*

The differences observed in growth location for *L. innocua* and both *L. lactis* strains can have a significant impact on their interactions as competitive species, and on the efficacy of nisin produced by *L. lactis* NZ9700, in particular its ability to reach *L. innocua* colonies. Additionally, the colony size and growth location can have a significant impact on the environmental stress response within a colony, i.e., self-induced acid stress, starvation/metabolic stress, oxidative stress, etc. Thus, significantly different responses to microbial decontamination processes are possible for systems of varying microstructure in the presence of natural microflora.

**Significance and Impact:**

Our findings give a systematic quantitative insight on the impact of co-culture with nisin-producing *L. lactis* on the growth of *Listeria*, in food model systems of varying structural complexity. They highlight the importance of accounting for bacterial stress adaptation in solid(like) systems when designing novel decontamination processes. Additionally, the microscopic differences observed using CLSM techniques are important to account for when designing such processes, as the colony size and location (both of the pathogen and the naturally present organism) directly affects the environmental stress within a colony, hence different survival rates of *Listeria* may be observed in different systems.

**Acknowledgements:**

This work was supported by the Department of Chemical and Process Engineering of the University of Surrey as well as an Impact Acceleration Grant (IAA-KN9149C) of the University of Surrey, an IAA-EPSRC Grant (RN0281J), and the Royal Society.

**References:**

[1] Valdramidis, V.P., Koutsoumanis, K.P., 2016. *Curr. Opin. Food Sci.* 12, 63–69.

[2] Troy, D.J. et al., 2016. *MESC* 120, 2–9.

[3] Gharsallaoui, A. et al., 2016. *Crit. Rev. Food Sci. Nutr.* 56, 1262–1274.

[4] Liao, H. et al., 2018. *Ultrason. Sonochem.* 42, 244–249.

[5] Li, W.-R. et al., 2010. *Appl. Microbiol. Biotechnol.* 85, 1115–1122.

[6] *The European Union summary report on trends and sources of zoonoses, zoonotic agents and food-borne outbreaks in 2014*, EFSA, 2015.

[7] Aspridou, Z. et al., 2014. *Food Res. Int.* 64, 683–691.

[8] Velliou, E.G. et al., 2013. *Food Res. Int.* 54, 1746–1752.

[9] Noriega, E. et al., 2013. *Food Microbiol.* 36, 355–364.

[10] Velliou, E.G. et al., 2011. *Food Microbiol.* 28, 702–711.

[11] Velliou, E.G. et al., 2010. *Lett. Appl. Microbiol.* 52, 116–122.

[12] Velliou, E.G. et al., 2012. *J. Therm. Biol.* 37, 72–78.

[13] Scatassa, M.L. et al., 2017. *Ital. J. Food Saf.* 6 (6191), 13-17.

[14] Baranyi, J., Roberts, T.A., 1994. *Int. J. Food Microbiol.* 23, 277–294.

[15] Geeraerd, A.H. et al., 2000. *Int. J. Food Microbiol.* 59, 185–209.

# 536267 Microdroplet-Enabled Metagenomic Reconstruction of Draft Genomes from the Human Gut Microbiome

**Abstract:**

The complexity of human gut microbiome and the "unculturable" majority of bacteria thwarts efforts to understand the mechanisms of the microbiome as a driver of human health. In response, the reconstruction of genomes from metagenomic datasets provides a means to access these "unculturable" members and link the taxonomy of individual members to functional capability. However, metagenomic reconstruction, without supplementation with other technologies or numerous samples over spatiotemporal instances, can only recover a small number of the most abundant members of complex communities or resolve very simple communities. To circumvent this, we developed a microfluidic approach to encapsulate fractionated human stool communities into microdroplets. This decreases the overall diversity and complexity in each sample. allowing access to rare microbiota from the microbial "dark matter" and facilitating the metagenomic reconstruction of their genomes. Using this approach, we demonstrate nearly complete reconstruction of draft genomes from a population co-cultivated in a microdroplet and provide insight into the functional and ecological potential of these microbes.

# 538118 Microdroplet Cocultivation and Characterization of Vaginal Bacteria in Vaginal Fluid

**Abstract:**

The human vaginal microbiome (HVM) plays a fundamental role in womenâ€™s health and susceptibility to sexually transmitted infections (STIs). For instance, bacterial vaginosis (BV) is characterized by the depletion of lactobacilli and an overgrowth of strict anaerobes. BV is associated with an increased risk of infertility, preterm birth, and acquiring STIs like HIV. Despite the importance of the HVM, the ecological roles of many vaginal species and their interactions within the microbiome as well as with the host remain unclear. In this work, we employ a microfluidic technology platform to encapsulate vaginal bacteria and cultivate them in nano-liter microdroplets. The small volume of these microdroplets lead to minimum requirement on reagents and enables us to make use of cervicovaginal fluid (CVF), a very limited resource that is used for cultivation. By incorporating CVF, our objective is to dissect bacterial inter-species interactions in the HVM under conditions closely mimicking the *in vivo* environment. We collect CVF from a cohort of donors of reproductive age and pool them for the droplet cultivation experiments. The CVF is used to cultivate several vaginal bacterial species, including *Lactobacillus iners,* the most frequently detected bacterium in the HVM, and *Gardnerella vaginalis,* a BV-associated bacterium. Monocultures and co-cultures are generated and characterized with cellular and molecular assays. From these experiments, we observe growth phenotypes that are different from those using laboratory media and reveal mechanisms that are relevant *in vivo*.

# 532045 ProbioticsÂ As Viable Antimicrobials Inhibiting Pathogens during Biofilm Formation

**Abstract:**

Biofilms are sessile microbial communities formed on the biotic and abiotic surfaces by the secretion of extracellular polymeric substances that enhance adherence to the surfaces and microbial aggregation. Biofilm bacteria are highly tolerant to exogenous stresses such as antibacterial agents due to mass transfer limitation in biofilm matrix and non-metabolizing nature of cells inside the biofilms. Many pathogens take advantages of living as biofilms, and thus 75%~80% of human infectious diseases are closely involved in the formation of biofilms. Therefore, understanding the biofilm formation is critical to establishing novel strategies of controlling infectious diseases. Although the majority of biofilm studies has examined mono-species cultures, the biofilm formation is intra- and inter-species phenomena that require dynamic interactions between bacteria in mixed biofilm communities. However, little is known to understand molecular interactions between biofilm bacteria. *Escherichia coli* strain Nissle 1917 (EcN) is a probiotic bacterium that has antagonistic effects on adherence, growth, and biofilm formation of other *E. coli* strains. In this study, we investigated the ability of EcN to outcompete with the biofilm formation of pathogenic bacteria such as pathogenic *E. coli* (EHEC), *Pseudomonas aeruginosa*, *Staphylococcus aureus*, and *S. epidermidis*. When dual-species biofilms were formed, EcN inhibited the EHEC biofilm population by 14-fold compared to EHEC single-species biofilms. This figure was 1,100-fold for *S. aureus* and 8,300-fold for *S. epidermidis*; however, EcN did not inhibit *P. aeruginosa* biofilms. In contrast, commensal *E. coli* did not exhibit any inhibitory effect toward other bacterial biofilms. We identified that EcN secretes DegP, a bifunctional (protease and chaperone) periplasmic protein, outside the cells and controls other biofilms. Although three *E. coli* strains tested in this study expressed *degP*, only the EcN strain secreted DegP outside the cells, and purified DegP directly repressed EHEC biofilm formation. Hence, probiotic *E. coli* outcompetes pathogenic biofilms via extracellular DegP activity during dual-species biofilm formation. In addition, through transposon mutagenesis screening of EcN, we identified that four genes *argC*, *argH*, *serB*, and *purL* are involved in repressing biofilms of EHEC as well as *S. aureus*. These results provide insights in applying and developing probiotics as an effective treatment for the biofilm-related infectious diseases.

# 532315 Microbiomes in Food, Health, and Bioprocessing: Advances and Challenges

**Abstract:**

Microbiomes are important to many ecosystems, including soils, guts, and bioprocessing (such as anaerobic digestion and wine fermentation). This talk will give an overview on recent advances in understanding microbiomes in several ecosystems and implications and applications to food, health and bioprocessing.

# 539207 Infusion of Walnut Husk into Polyethylene

**Abstract:**

Black walnuts, juglans nigra, are indigenous to eastern North America, and the nut is composed of the kernel, shell, and husk. In commercial production, the nuts are harvested from the ground and sold to processors that de-husk the walnut and the de-husked walnuts are then shipped for further processing. The green, or fresh, husk is commonly discarded. Green walnut husks contain a wealth of polyphenolic compounds, tannins, and other chemicals, and the husk exhibits antimicrobial and antioxidant properties. Walnut husks may also be used as a bio-herbicide. These antimicrobial compounds may be extracted from black walnut husk using supercritical carbon dioxide with an ethanol modifier, with the resulting extract exhibiting antioxidant and antimicrobial properties. Additionally, supercritical carbon dioxide can cause polymers to swell, thereby allowing polymers to be infused with chemicals while exposed to supercritical carbon dioxide. It will be presented that carbon dioxide and ethanol can be used in extracting chemicals, as determined by HPLC-MS analysis, from walnut husk while simultaneously infusing chemicals into polyethylene. The resulting extract showed antioxidant potential as measured by the total phenolic content (TPC) assay and antimicrobial properties. The treated polyethylene exhibited antimicrobial effects compared to an untreated control. The treated polyethylene was also analyzed by differential scanning calorimetry (DSC) and thermal gravimetric analysis (TGA) and compared against a control. Additionally the treated polyethylene was then heated to its melt temperature and its antimicrobial properties re-evaluated.

# 536082 Implementation of a Mixed Integer Linear Programming Approach to Establish *De Novo* Synthesis Routes of Antioxidants Derived from the Fermentation of *Theobroma Cacao* Seeds

**Abstract:**

INTRODUCTION

Cacao (*Theobroma cacao*) is a plant of South American origin rich in nutrient contents widely commercialized around the globe because of the production of chocolate. Chocolate is a highly demanded product due to its organoleptic characteristics and nutritional value. In order to obtain this product, cacao seeds covered by a mucilaginous pulp of the fruit are fermented in boxes in a traditional way. This process lasts about 6 to 7 days and begins whit the pulp degradation and consequent formation of the flavor, odor and color precursors which, strongly influence the chocolateâ€™s quality. Three main types of microorganisms found in the fermentation process have been proven to be essential to obtain of high-quality chocolate: yeasts, acetic and lactic acid bacteria. Among the nutritional compounds and organoleptic properties than make cacao a very desirable product with a high added value, the most important are: antioxidants and astringent, bitter and sweet flavors. Considering the nutritional and economic importance of the compounds found in fermented cacao seeds, a need to deeply understand and determine the metabolic routes associated with the biochemical reactions they are involved in arises. Then this work aimed to propose novel routes to synthesize antioxidants derived from enzymes present in the fermentation process of Theobroma cacao seeds that could be used to construct a complete process of production in bioreactor using heterologous expression. To gain in the knowledge of those products, it is proposed to create a catalogue of the metabolic routes associated with the synthesis of both nutritional and organoleptic compounds.

METHODS

First, we determined the optimal reactant and product combination with the overall stoichiometry of overall reactions proposed for obtain the naringenin flavonoid; a. from glucose, b. the amino acid L-tyrosine and c. L-phenylalanine and d. shikimate, to obtain the maximum achievable yield for all the possible bioconversions. This was carried out considering the thermodynamic restrictions. Metabolites and reactions required database were obtained based on reported metabolic reconstructions of individual species involved of the fermentation process and the complete KEGG database was also used. All reactions were considered reversible in the databases. Substrates and products were obtained using the polyphenol composition of the seed and an antioxidant predictor respectively. Then, it was identified the routes and reactions that satisfy the identified overall stoichiometry taking into account the free energy change of the reactions and minimizing the amount of reactions required to the pathways. Subsequently, general reactions were generated to produce naringenin compounds, and the catalog of metabolic pathways associated with the synthesis of nutritional and organoleptic compounds of fermentation were prepared.

RESULTS

By means of both the algorithm and the catalogue, eleven metabolic routes for each of the proposed general reactions that are involved in the synthesis of the most relevant antioxidants were obtained from glucose, L-tyrosine, L-phenylalanine and shikimate, and later, from naringenin. The catalogue made easier the search for compounds involved not only in the fermentation process of cacao, but in any other fermentation process in food industry.

CONCLUSION

Due to the economic and nutritional importance of chocolate, the identification of the routes of synthesis of precursor compounds, organoleptic and nutritional characteristics, helps in the design of high quality products, not only in the chocolate industry but also in any other product Fermented industry.

# 533731 Encapsulation of Lactic Acid Bacteria By Multiple Emulsion System

**Abstract:**

In this study, we prepared functional food powder by multiple emulsions, which contained lactic acid bacteria with spray drying process. Generally, the functional food with the probiotic bacteria provides beneficial health effects, especially for digestive system. However, LAB has sensitive characteristics and is liable to damage under harsh and detrimental conditions no matter it exists in human digestive system or circulation storage. In order to maintain the viability of probiotic bacteria during processing and storage afterwards, encapsulation of probiotic bacteria by multiple emulsions was introduced for development of value-added product. Multiple emulsions are really complex with both coexistence of water-in-oil (W/O) and oil-in-water (O/W). They can be classified into two major types: water-oil-water emulsion (W/O/W) and oil-water-oil (O/W/O). In this study, W/O/W emulsion type was prepared for consideration of application potency in the future. Thanks to the multi-compartmentalized structure, multiple emulsions can prevent the functional components from environmental damage as result of masking flavor or releasing control. So as to meet the above requirements, we selected higher heat-tolerant lactic acid bacteria as encapsulated target through foodâ€“grade W/O/W multiple emulsions by a two-stage process.

To design the targeting formulation, edible proteins and polysaccharides were used in extra water phase as excipients. The survival rate of lactic acid bacteria and recovery efficiency were evaluated during the process. The survival rate of encapsulated lactic acid bacteria with stimulation of gastrointestinal conditions was also studied. The results clearly showed that the viable cell count of multiple emulsions had a marked increase along with formulation design. The viable probiotic cells obtained from emulsion was around 5.8Ã—$10^8$ CFU/mL and 8.1Ã—$10^7$ CFU/g after spray drying. The decreased amount of encapsulated lactic acid bacteria was less than 1 order of magnitude for viability assay under simulated gastrointestinal conditions. On the contrary, the free cells showed a higher decrease in their viability under same conditions (> 5 log CFU/g). The results obtained in this study suggested that multiple emulsions could play as a good delivery system and provide good protection for lactic acid bacteria and extend its chemical stability.

# 531758 Semi-Continuous Fermentation of Acetic Acid By Mutant of *Acetobacter Pasteurianus*


**Abstract:**

Acetic acid, as the main component of vinegar, is used as acidifying agent, flavoring agent and spice in the food industry. It is widely used in the production of anhydride acetic acid, vinyl acetate, acetate fiber, metal acetate, acetic ester and other chemicals. Although there are many methods of producing acetic acid, the biological method is the only method to produce acetic acid used in the food industry. *Acetobacter pasteurianus* is the main species used in vinegar fermentation in China. However, the yield by them is not high, and the stain is unstable. In order to improve the yield of acetic acid, in this study, a mutant *Acetobacter pasteurianus* YHL3 screened by UV from *Acetobacter pasteurianus* CICC 20001 was selected. Its semi-continuous fermentation process was optimized. With the optimized medium, the highest acetic acid production of 75.24 g/L was obtained in a 7-L bioreactor in four repeated batches.

The mutagenic mechanism of *Acetobacter pasteurianus* YHL3 was also analyzed. The tolerance test results with YHL3 and CICC 20001 showed that with the high alcohol content, ADH and ALDH with YHL3 were 5.96±0.26 U/mg and 4.82±0.22 U/mg, respectively, which were significantly higher than those of CICC 20001 (4.78±0.28 U/mg and 4.12±0.24 U/mg, respectively). With the high acid, ADH and ALDH with YHL3 were 6.08±0.18 U/mg and 5.47±0.16 U/mg, respectively. Which were also higher than those of CICC 20001 (5.66±0.16 U/mg and 4.87±0.12 U/mg, respectively). According to the metabolic pathway network analysis of *Acetobacter pasteuriens*, it was found that the enzyme system of YHL3 in ethanol oxidative respiratory chain was better than that of CICC 20001. The above results demonstrated the feasibility of industrial production of high-concentration acetic acid in vinegar fermentation.

# 532342 Production of Heavy Oil Liamocin By *Aureobasidium Pullulans*

**Abstract:**

*Aureobasidium pullulans* is a ubiquitous fungus that is well known for producing exopolysaccharide pullulan, polymalic acid (PMA), and many useful industrial enzymes such as xylanase and lipase. Some strains of *A. pullulans* can also produce extracellular heavier-than-water oils known as liamocins. Liamocins are esters composed of a single mannitol headgroup O-acetylated with 3 or 4 3,5-dihydroxydecanoic ester groups, and can be used as biosurfactant, antimicrobial agent against some *Streptococcus* spp., anti-cancer agent, and biofuels. The highest liamocin production reported to date was 22 g liamocin/L from optimized medium by strain NRRL 50384. The PKS (polyketide synthase) is reported to be involved in the biosyntheses of fatty acids and melanin in *A. pullulans*. However, the complete liamocin biosynthesis pathway has not been elucidated, although the genomes of several strains have been sequenced.

The objective of this research was to optimize liamocins production from various sugars, such as sucrose and glucose, in fermentation. The strain NRRL 58544 produced 9.2 g/L liamocins from 50 g/L sucrose, which was higher than 5.4 g/L previously reported using the same medium formula. Batch fermentations in shake-flasks showed that a higher liamocin production was obtained in the medium with a higher C/N ratio. Furthermore, a lower temperature stimulated oils accumulation. In general, dark/black heavy oils were produced from sucrose, because of the coproduction of melanin. In contrast, heavy oils produced from glucose by the strain NRRL 62031 had a more pleasant, light yellow color. Scale up of the fermentation process will be performed in a 1-L bioreactor with pH and temperature controls. Transcriptomics and metabolomics analyses of the fermentation under different conditions will be performed, and the results will provide a theoretical basis for genetic modifications of the metabolic pathway to further enhance liamocin production in *A. pullulans*.

# 528009 (Keynote) Efficient Biosynthesis of Omega-3 PUFA: From Lab to Factory

**Abstract:**

As one of the three major nutrients, the consumption of lipid quality and quantity is closely related to the occurrence and development of lifestyle-related diseases chronic disease, such as cardiovascular and cerebrovascular diseases, obesity, and type 2 diabetes. WHO reported that dietary and nutritional factors are more effective approaches on health than medical factors. Omega-3 fatty acids, especiallay eicosapentaenoic acid (EPA) and docosahexaenoic acid (DHA), are critical for human health. It is estimated that human should have an omega-6:3 PUFA ratio of 1:1. However, current Chinese diet have an imbalance omega-6:3 ratios of 20 or even high. Marine fish are traditionally known as a natural source of EPA and DHA. However, the limitations of this source, such as seasonal availability, the ecological impact of fishing and the resulting environmental pollution, make it difficult to fulfill the nutritional requirements of human beings. In the past years, we have been conducting research work to engineer a sustainable land-based microalgae source of EPA and DHA to reduce pressure on fish stocks to meeting future demand. We isolated a genetic stability DHA producer industrial strain *Schizochytrium* sp. by high throughput screening method and the metabolic pathway analysis. Then, multi-index feedback cycle fermentation control technology and new bioreactors were developed to enhance lipid production. In parallel, a novel green lipid extraction technology has been developed that contains super-high levels of extraction, in the range of 90%-95%. These developments will provide theoretical basis and technological support for the industrial cultivation of oil-producing microorganism.

# 541778 Food-Energy-Water Nexus Systems Engineering

**Abstract:**

Due to population growth, economic development, urbanization, climate change, and natural resource degradation, the global demand for food, energy and clean/freshwater is rapidly increasing [1]. As demand grows the stresses and interdependences between these three resource systems, which are commonly referred to as Food-Energy-Water Nexus (FEW-N), are becoming more apparent [2,3]. Decision making can be very challenging due to multiple dimensions of the biophysical water, energy and food systems, and the players and stakeholders connected with them [4]. A holistic systems engineering approach is thus clearly needed to navigate the multi-faceted FEW-N space, identify opportunities for synergistic benefits and systematically explore interactions and trade-offs [5].

In this work, we present the foundations of a systems engineering framework and quantitative decision-making tools for the analysis and trade-off optimization of stressed interconnected resource FEW-N networks. The framework combines data analytics and mixed-integer modeling and optimization methods establishing the interdependencies and potentially competing interests amongst the food, energy and water elements in the system, along with policy, sustainability and feedback from the various stakeholders. A multi-objective optimization strategy is followed for the analysis of the trade-offs empowered by the introduction of composite FEW-N metrics as means to facilitate decision making and compare alternative process and technological options. The versatility, potential and applicability of the proposed framework is demonstrated by three nexus case studies, (i) a complex power generating system in the South-Central Texas region [6], (ii) a dairy production and processing plant [7], and (iii) a land use allocation problem on a regional farming system in China [8].

**References**

[1] D'Odorico, P., Davis, K. F., Rosa, L., Carr, J. A., Chiarelli, D., Dell'Angelo, J., ... & Rulli, M. C. (2018). The Global Foodâ€Energyâ€Water Nexus. Reviews of Geophysics.

[2] McCarl, B. A., Yang, Y., Schwabe, K., Engel, B. A., Mondal, A. H., Ringler, C., & Pistikopoulos, E. N. (2017). Model Use in WEF Nexus Analysis: a Review of Issues. *Current Sustainable/Renewable Energy Reports*, *4*(3), 144-152.

[3] McCarl, B. A., Yang, Y., Srinivasan, R., Pistikopoulos, E. N., & Mohtar, R. H. (2017). Data for WEF Nexus Analysis: a Review of Issues. *Current Sustainable/Renewable Energy Reports*, *4*(3), 137-143.

[4] Daher, B., Mohtar, R.H., Pistikopoulos, E.N., Portney, K.E., Kaiser, R. and Saad, W. (2018) Developing Socio-Techno-Economic-Political (STEP) Solutions for Addressing Resource Nexus Hotspots, Sustainability 2018, 10, 512; doi:10.3390/su10020512

[5] Garcia, D. J., & You, F. (2016). The water-energy-food nexus and process systems engineering: a new focus. *Computers & Chemical Engineering*, 91, 49-67.

[6] Diangelakis, N. A., & Pistikopoulos, E. N. (2017). A multi-scale energy systems engineering approach to residential combined heat and power systems. *Computers & Chemical Engineering*, 102, 128-138.

[7] Avraamidou, S., Milhorn, A., Sarwar, O., & Pistikopoulos, E. N. (2018). Towards a quantitative Food-Energy-Water Nexus metric to facilitate decision making in process systems: A case study on a dairy production plant. *28th European Symposium on Computer-Aided Process Engineering*: Accepted.

[8] Nie, Y., Avraamidou, S., Li, J., Xiao, X., & Pistikopoulos, E. N. (2018). Land use modeling and optimization based on food-energy-water nexus: a case study on crop-livestock systems. *13th International Symposium on Process Systems Engineering*: Accepted.

***Renewable Carbons from Food Waste for Separation and Catalysis Technologies***

This study explores the production of renewable and sustainable carbons from waste biomass resources, for adsorption and catalytic applications. Specifically, the research will explore the production of carbons derived from **food waste** and their utilization as sorbents of aromatic contaminants in **water** and catalyst supports for the production of clean **energy and fuels**.

The idea of using renewable and sustainable resources to replace the carbon derived from fossil fuels is not new. Researchers have been working on the utilization of various biomass resources, such as rice husk, coconut shells and others for years.[1–5] However, the utilization of food waste as a carbon resource has not been extensively explored. Food waste could be the ultimate resource for carbon production, since it is produced in excessive volumes and it represents one of the most significant environmental problems, which is urgently looking for a solution. The critical question is whether the activated carbons derived by food waste have the potential to be used for purification and/or catalysis purposes. Food waste is heterogeneous in nature and contains various types of minerals, which may provide advantages or disadvantages, depending on the application.

In our lab, we receive the food waste from the Dining Halls of the University of Connecticut and after pretreatment, that includes washing, grinding and sieving, we perform pyrolysis and activation to turn the waste into carbons. The produced carbons are then characterized and they are tested as sorbents and catalyst supports. Our goal is to understand how the conditions of pyrolysis and activation affect the properties of the carbons and, consequently, their effectiveness as sorbents and catalyst supports. During the pyrolysis reaction, temperature and residence time are the most important parameters that affect the yield and the properties of the produced biochar. Activation of the biochar is performed either using steam, physical activation, or using chemicals (e.g. $H_3PO_4$), chemical activation. During the former method of activation, temperature and residence time are, again, among the most important parameters that affect the surface area, the carbon content and the porosity of the produced carbons. During the latter method, the impregnation ratio between the chemical and the biochar is the most critical.

During this talk we will discuss the properties of the activated carbons produced by food waste and how these properties are affected by the reaction conditions and the method of activation. Surface area, porosity, carbon content, surface functional groups, morphologies are the properties of interest which will be discussed. Then, we will discuss the potential of using the activated carbons produced by food waste, as sorbents of aromatic hydrocarbons in water and catalyst supports for the hydrodeoxygenation of bio-oil.

**References**

(1) Tripathi, M.; Sahu, J. N.; Ganesan, P. Effect of Process Parameters on Production of Biochar from Biomass Waste through Pyrolysis: A Review. *Renew. Sustain. Energy Rev.* **2016**, *55*, 467–481.

(2) Kwapinski, W.; Byrne, C. M. P.; Kryachko, E.; Wolfram, P.; Adley, C.; Leahy, J. J.; Novotny, E. H.; Hayes, M. H. B. Biochar from Biomass and Waste. *Waste and Biomass Valorization* **2010**, *1* (2), 177–189.

(3) Ahmad, M.; Rajapaksha, A. U.; Lim, J. E.; Zhang, M.; Bolan, N.; Mohan, D.; Vithanage, M.; Lee, S. S.; Ok, Y. S. Biochar as a Sorbent for Contaminant Management in Soil and Water: A Review. *Chemosphere* **2014**, *99* (November 2013), 19–23.

(4) Tan, Z.; Lin, C. S. K.; Ji, X.; Rainey, T. J. Returning Biochar to Fields: A Review. *Appl. Soil Ecol.* **2017**, *116* (September 2016), 1–11.

(5) Azargohar, R.; Dalai, A. K. Biochar as a Precursor of Activated Carbon. *Appl. Biochem. Biotechnol.* **2006**, *131* (1–3), 762–773.

# 541779 A Computational Framework for Sustainable Waste Management and Simultaneous Recovery of Nutrients and Energy

**Abstract:**

To protect human health and the environment, communities should have sustainable infrastructures to manage waste. Critically, livestock waste may cause some air quality degradation from ammonia and methane emissions, soil quality detriment due to in-excess nutrients and acidification, and water pollution issues resulting from nutrient and pathogens runoff to water bodies, which leads to eutrophication, hypoxia, and harmful algal blooms (HABs).

In addition, the recovery of carbon from organic waste offers an alternative for replacing fossil fuels with renewable sources for energy production. Despite the significant environmental benefits by performing pollution management of these organic materials, the deployment of technologies for obtaining value-added products and renewable energy from livestock waste is not a current practice due to the high investment costs and low product market values.

This contribution describes a multi-scale approach for the development of a supply chain framework to identify and evaluate techniques and logistics for the management of solid waste generated from primary sources of nutrient pollution. This effort will inform stakeholdersâ€™ waste management options while assessing potential economic and environmental benefits. Other potential benefits include the reduction of nutrient pollution and HABs, recovery of valuable products, and adoption of nutrient-reducing technologies in different locations (e.g., community, state, regional, etc.). In addition, the proposed framework captures techno-economic and logistical issues and can accommodate diverse types of material management incentives obtained at federal and state levels.

# 542591 Electrochemical Conversion of Ammonia and Nitrogen for Sustainable Food-Energy-Water

**Abstract:**

**Electrochemical Conversion of Ammonia and Nitrogen for Sustainable Food-Energy-Water**

Gerardine G. Botte

Center for Electrochemical Engineering Research

Department of Chemical and Biomolecular Engineering

Russ College of Engineering and Technology

Ohio University, Athens

OH 45701

The production of reactive nitrogen fertilizer via the Haber-Bosch (HB) process, combined with industrialization and intensification of agriculture, has created an abundant global food supply to support population growth, but has doubled the amount of reactive nitrogen compared to pre-industrial levels.[1] With those perspectives, ammonia can be seen as an important chemical of use or as an environmental challenge. Ammonia emissions into air (ambient ammonia) and water represent an environmental challenge. Ambient ammonia not only contributes to inorganic PM2.5 (particulate matter with an aerodynamic diameter of less than 2.5 Î¼ m) directly but also plays an important role in secondary organic aerosol formation by interacting with gaseous phase organic acids and forming condensable salts. In addition, ammonia emissions in water are associated with environmental problems such as algae bloom. Various industries and other operations are considered ammonia emitters, but the ones related to sustainable food production such as livestock management and fertilizer application are among the major contributors accounting for 60-75% of the ammonia emissions.[2-3]

On the other hand, ammonia produced commercially via the HB process is usually known as one of the greatest inventions of the 20th Century. The HB process made it possible for the first time to produce synthetic fertilizers and produce sufficient food for the Earthâ€™s growing population. However, the HB process requires high temperatures and pressures and as a result it can only be done in large plants which consume large amounts of power and generate large quantities of $CO_2$ from the fuels used to power the process, such as natural gas. It is estimated that the HB process accounts for nearly 1% of the entire global power consumption. Due to the sheer scale of the HB process, distribution from the point of production to the point of use becomes an additional carbon and energy burden. Alternative methods of manufacturing ammonia that enable small scale, distributed generation at the point of use using renewable energy and sustainable feedstocks could have a great impact on global $CO_2$ emissions.

A closed loop approach that includes producing fertilizer at the point of use, applying nitrogen fertilizer as close to the time that crops needs it, and recovering and recycling nitrogen fertilizer can contribute to address the challenges related to ammonia emissions, while decreasing water and energy consumption. Electrochemical conversion of nitrogen and ammonia presents an excellent platform to achieve this closed loop approach. Dr. Botte and members of her research group at the Center for Electrochemical Engineering Research (CEER) are working on the electrochemical conversion[4]/synthesis of ammonia[5] in alkaline media. In this talk, recent advances in the platform process for ammonia and nitrogen removal/recycling and ammonia synthesis will be discussed as well as their application for sustainable food, energy, and water.

References:

1. Grand Challenges - Manage the Nitrogen Cycle [WWW Document], n.d. URL http://www.engineeringchallenges.org/9132.aspx (accessed 5.28.18).

2. G. E. Mansell, "An Improved Ammonia Inventory for the WRAP Domain (Final Report)", ENVIRON International Corporation, Report No. 415.899.0700, March 7 (2005).

3. FAO. Global estimates of gaseous emissions of $NH_3$, NO, and $N_2O$ from agricultural land, International Fertilizer, Industry Association. Rome. 2001.

4. Botte, G. G., Cooper, M. and Vitse, F., (2009) Electro-catalysts for the oxidation of ammonia in alkaline Media, U.S. Patent No. 7,485,211.

5. B. L. Sheets, G. G. Botte, Electrochemical nitrogen reduction to ammonia under mild conditions enabled by a polymer gel electrolyte, Chemical Communications 54 (2018), 4250-4253.

# 542589 Encapsulation and Nanoparticle Formation for “Non-Standard” Applications

**Abstract:**

Encapsulation and nanoparticle formation for “non-standard” applications

Prof. Robert K. Prud'homme[1], Prof. Rodney D. Priestley[1], Leslie Shor[2] ,Douglas Scott[1], Jie Feng[1]

[1] Princeton University, Department of Chemical and Biological Engineering, Princeton University, Princeton NJ 08544

[2] University of Connecticut, Department of Chemical and Biomolecular Engineering, Storrs, CT 06269-3222

Presenting author: Robert K. Prud'homme, prudhomm@princeton.edu

Traditionally, encapsulation and nanoparticles have been associated with high value applications such as drug delivery in medicine. The two main reasons have been the high cost of nanoparticle assembly processes for complex nanoparticles that encapsulate and release actives in a controlled way, and the fact that encapsulation technologies for larger scale macro-particles do not translate to nano-scale particles. Recent advances in low-cost nanoparticle assembly present the opportunity to open up new applications. First, we present an overview of encapsulation and release technologies. Then we provide a review of the rapid-precipitation technologies we have developed that rely on the control of the fluid dynamics, diffusion-limited aggregation, and steric stabilization of nanoparticles. The processes are scaleable, low cost, and produce nanoparticles from 50-400 nm. Several novel applications will be presented: (1) nanoparticle delivery of agricultural actives s to plant roots by protist active transport, (2) triggered indicator inks for sterilization processes, (3) “green chemistry” encapsulation of personal care compounds using purely aqueous processing.

# 534657 Hybrid Mixture Theory-Based Modeling of Moisture Transport in Carrots during Drying

**Abstract:**

It is necessary to understand the water transport phenomenon occurring in a food product since it plays an important role by affecting its functional properties during processes. It influences the appearance, structure, texture, and ultimately consumer acceptance of a product. Proposed models in the literature have primarily used Fickian type diffusion to describe the water transport in biopolymers. These models fail to include the effect of glass transition and relaxation of the structure during adsorption/desorption on water transport and swelling. Therefore, in this study, the moisture transport mechanism during drying process was elucidated for carrots using numerical solution of hybrid mixture theory (HMT) based multi-scale fluid transport equations, which include the stress relaxation terms. The drying experiments were performed in an environmental chamber (Associated Environmental Systems, Ayer, MA, USA) controlling the temperature and the relative humidity of the environment for drying of carrots up to 4 hours. The stress relaxation data, required for the transport equations, was obtained using a dynamic mechanical analyzer (DMA). The three-element Maxwell model was fitted well to describe the viscoelastic behavior of carrots ($R^2$>0.999, RMSE <$2.08\times10^{-4}$). Since drying causes shrinkage throughout the sample, the transport equations were transformed from the Eulerian coordinates to the stationary Lagrangian coordinates. Numerical simulations gave data on temporal and spatial profiles for moisture, viscoelastic stress distribution, and shrinkage. Good agreements between the experimental and the predicted values of moisture content and volume changes were obtained. During simulations, the center of the carrots was observed as the region with the highest moisture content while the boundaries presented the lowest values. Also, the HMT based non-Fickian transport model was coupled with kinetic equations representing the quality attributes of carrots, such as color and texture to predict these attributes as a function of drying time. The results of the color parameters (L, a, and b), showing decreasing trends between 30.12-51.58% after 4 h drying, were compared with the model predictions. In overall, the created model provided significant information to fundamentally describe the underlying mechanisms that occur during the drying process. The simulations showed that the moisture transfer, the shrinkage effect, and the quality attributes can be predicted with reasonable accuracy. This model would allow determining the optimum drying conditions to obtain improved product quality and prolongation of the shelf-life for food products like carrots.

# 542730 Separation of Chitin from Shrimp Shells using Functional Ionic Liquids

**Abstract:**

Chitin, the liner chain of (1, 4)-linked-2-acetylamido-2-deoxy-Î²-D-glucose, is the second most abundant natural polymer mostly from shrimp and crab shells in the industry.[1] Chitin and its derivates are biodegradable, nontoxic and chemical stable, which have been widely applied in the medicine, agriculture, textile industries and other related fields.[2] The traditional method of preparing chitin needs repeated acid and alkali treatment to remove protein and calcium carbonate and high-water consumption to wash the products. Obviously, the method leads to serious environment pollution and large water cost. Several approaches, such as ionic liquids extraction, enzymatic reaction or microbial fermentation, have been reported to solve above problems. Those approaches still have defects of the high cost, low purity and long time. Hence, it is meaningful to develop an efficient, simple and green strategy of chitin synthesis from shrimp shells.

Herein, we report a green and simple method to prepare chitin using choline-based ionic liquids (ILs) from shrimp shells in one step. The used IL could remove the protein and calcium carbonate, simultaneously. The type of the IL, experiment temperature and time played key roles in the process. Under optimal conditions, the chitin with 92 % purity was obtained, and the yield was up to 78 %. The reason for calcium carbonate and protein removal was the different roles of the cation and anion in the used IL. The proposed method provides a promising alternation of chitin preparation in the industry.

**Key words:** shrimp shells, chitin, ionic liquids

**Acknowledgments**

This work was supported financially by National Natural Scientific Fund of China (No.21476234, No.21506231).

**References**

[1] G. Cárdenas, G. Cabrera, E. Taboada, S. P. Miranda, Journal of Applied Polymer Science 2004, 93: 1876-1885.

[2] C. E. Blank, R. W. Parks, N. W. Hinman, Journal of Applied Phycology 2016, 28: 2753-2766.

# 539409 *Assessment of Oxidative Stability of Home-Cooked Meat Products in US By Targeted Lipidomics*

**Abstract:**

***Assessment of oxidative stability of home-cooked meat products in US by targeted lipidomics***

*Lisaura Maldonado-Pereira[1], Matthew Schweiss,[2] Ilce G. Medina-Meza PhD[1,2]*,*

*maldon60@msu.edu, schweis3@msu.edu ilce@anr.msu.edu*

*[1]Chemical Engineering and Material Science Department*

*[2]Biosystems and Agricultural Engineering Department,*

***Keywords: Lipid oxidation, food toxicology, lipidomics,***

**Abstract**

Lipid oxidation is one of the most important factors limiting the shelf life and commercial stability of meat and meat products. Meat oxidation is related to the content of natural antioxidants and the polyunsaturation degree of fatty acids.

Monitoring and reducing lipid oxidation in food has become a relevant topic in the meat industry due to the strict regulations of food quality products and its effect in human health. Lipid oxidation is a significant problem relative to off-flavour and off-odor. Warmed-over flavor appears to be related to lipid oxidation in meat and meat products. Oxidation can occur in either the stored triglycerides or the tissue phospholipids. Ferric heme pigments have been implicated as the major pro-oxidants in tissue lipid oxidation. Pigment and lipid oxidation are interrelated, and ferric hemes are believed to promote lipid oxidation. Oxidative deterioration of lipids directly affects a number of quality characteristics in meat and meat products, including flavor, color, texture, nutritive value and safety. It has been widely accepted that lipid oxidation in food occurs via a free radical chain mechanism that proceeds through three distinct stages of initiation, propagation, and termination, leading to a series of complex chemical changes.

Cooked meat undergoes rapid deterioration due to tissue lipid oxidation. The meat pigment in the cured pink ferrous form does not promote the rapid oxidation undergone by cooked uncured meat. For example, poultry meat is characterized by high protein content and a low-fat content (around 20 g and 5 g fat/100 g raw meat without skin), respectively, however chicken lipids display a high level of unsaturated fatty acids, which are usually perceived as â €˜healthyâ€™ by consumers, especially the polyunsaturated ones. In addition, lipid oxidation and myoglobin oxidation in meat lead to off-flavor development and discoloration. These processes appear to be linked and the oxidation of one of these leads to the formation of chemical species that can exacerbate oxidation of the other.

Evaluating lipid oxidation status is a challenging task due to a number of facts like the different compounds which are formed depending on the time, the extent of oxidation and the mechanism involved. Therefore, choosing just one parameter to analyze the oxidative status is rather difficult and it is frequently more convenient to combine different methods.

In addition, different studies have found a direct relationship between increased quantities of lipid oxidation products and food processing conditions (i.e. temperature, cooking time, and packaging methods). Besides the effect of lipid oxidation in meat quality, it has been proved that there is correlation between lipid oxidation products (derived from the oxidation process of cholesterol) and chronic diseases, such as cancer, atherosclerosis, and even neurological diseases such as Parkinson and Alzheimer. This suggests a potential risk in long-term consumption of these lipid oxidation products. This suggests the existence of a variability in lipid oxidation products consumption per diet. Thus, the aim of this study was to evaluate the influence of pan frying cooking method on the lipid oxidation status of the most consumed meats and meat products in the USA diet by a lipidomic approach. Primary and secondary oxidation products were measured on chicken, pork chop, ground beef 80% lean, ground beef 93% lean by Peroxide Value (POV) and Thiobarbituric acid (TBARs), respectively. Additionally, quantification of Fatty Acids Methyl Esther (FAME) was performed by Gas Chromatography (GC) means. Raw and cooked meat products were evaluated, as well as, other oxidative derivatives, such as Cholesterol Oxidation Products (COPs) by GC-MS and LC/MS/MS.

Results show that pan frying has a higher effect on the formation of primary oxidation products in pork chop with a decrease of 18% on POV value, in comparison with raw samples (18.60 meq $O_2$/kg lipid). As expected, the pan frying method has also a higher effect on the formation of secondary oxidation products in pork chop. TBARS shown an increase of 30% mg MDA/kg lipid between raw (553.05 mg MDA/kg lipid) and cooked samples. (1795.03 mg MDA/kg lipid). After FAME quantification, values for saturated fatty acids percentages (SFA%) ranges between 47.32 to 87.36 (%w/w) in raw samples and 38.99 to 89.28 (%w/w) in cooked samples. Monounsaturated fatty acid percentages (MUFA%) range between 6.27 to 51.02 (%w/w) in raw samples and 10.18-55.54 (%w/w) in cooked samples. Polyunsaturated fatty acids percentages (PUFA%) range between 1.31 to 9.35 (%w/w) in raw samples and 0.54 to 5.48 (%w/w) in cooked samples. Quantification of COPs and other derivatives was as well performed. These results follow the proposed mechanism of chain reaction that states the formation of primary oxidation products (peroxides) that usually suffer further oxidation turning into secondary oxidation products such as aldehydes, ketones, epoxides. Further studies are needed to fully characterize oxidation products and understand its influence in both meat quality and health related to its long-term consumption.

# 534943 Process Development for the Spray Drying of Milk Protein Stabilized Emulsions with High Oil Content

**Abstract:**

Spray drying is one of the most cost effective and scalable method for ingredient microencapsulation. Lipophilic ingredients, such as fish oil, beta carotene, and hydrophobic minerals and vitamins, can be encapsulated with carbohydrates as wall materials. We present a process development study on the spray drying of milk protein stabilized emulsions with high oil content. First, we found that pre-evaporation of raw milk not only reduces energy cost and equipment space occupation, but also improves the stability of pre-spray emulsions and thus reduces fat loss in this process. Second, adding carbohydrates provides a stronger barrier that reduces the portion of surface extractable fat, a main cause of oil leakage during long-term storage particularly for high oil content powder. The hydrophilicity of carbohydrates also improves powder dispersibility through dissolution-enhanced wetting. We found that carbohydrates with smaller molecular weights perform better at reducing surface extractable fat and improving surface wettability. This product will serve as the main ingredient of a powdered ready-to-use therapeutic food to treat severe malnutrition in India.

# 522788 Linking Metabolizable Energy to Chemical Oxygen Demand

**Abstract:**

A human and the humanâ€™s intestinal microbiome interact in complex ways that influence human health. Interactions involving metabolism are very pronounced and have captured much attention. To deepen understanding of human-microbe metabolic interactions, research experts in human metabolism, microbial metabolism and ecology, and engineering are joining forces. For historical and topical reasons, these disciplines use different metrics to quantify mass and energy. The metabolizable energy (ME, units: kcal) of food is a primary metric for human nutrition and metabolism. ME is the net energy within the food that is available to humans after digestion; it is equivalent to the gross energy intake minus body wastes (i.e., urine, feces, gases, and shedding). Microbial ecologists and environmental engineers track the energy embodied in electron equivalents using the chemical oxygen demand (COD). COD is the amount of oxygen required to oxidize organic matter, and it is related to electron equivalents via 8 g COD/ $e^-$ eq. In order to develop a consistent means to link energy and electron balances, which is essential for integrating the roles of human and microbial metabolism, an essential first step is to link ME and COD.

The purpose of this study was to develop a correlation between COD and energy values in common food items. We developed a framework to convert between COD and ME in four steps. First, we gathered food items representing the three macronutrients -- carbohydrates, protein, and fat â€“ and classified them according to their complexity: monomers, polymers, or complex ingredients. Using the carbohydrate macronutrient as an example, glucose is classified as a monomer, resistant starch as a polymer, and maple syrup as a complex ingredient. Second, we developed a chemical formula of each food item to calculate the theoretical COD. The monomers and polymers have known formulas. We calculated the complex ingredientsâ€™ formulas from nutrient data on the USDA food database: The macronutrient composition of the food items listed in the database can be used to determine the molar ratio of carbon, hydrogen, oxygen, and nitrogen atoms to derive a molecular formula. Third, we compared the theoretical COD with the experimental COD values measured using a commercial COD measurement kit. For the complex ingredients, the measured COD was used to confirm accuracy of the developed molecular formula. Fourth, the ME of the food was calculated using the Atwater equations. The result of this framework is a value for the ME and the COD for each of the food items. This methodology was completed for representative food items. The results were used to establish a correlation between ME and COD values.

Our results show that the developed molecular formulas accurately predicted the measured COD. For example, the theoretical and measured COD of egg whites agreed within ±1%. When the COD values were compared with the energy values, the best fit was linear. We observed an $R^2$ value calculated using the Pearson correlation of 0.992, a slope of 3.4 kcal/gCOD, and an intercept of 0.15 kcal/g. These results demonstrate that accurate conversion between ME and COD is possible for many food items. By providing a rubric for linking energy (kcal) measurements with electron equivalents (COD), we establish a framework to connect the bioenergetic interactions of humans and their microbiomes. This is an essential first step towards integrating microbial and human mathematical models and interpreting clinical results.

# 538967 An Experimental and Computational Study of Saponin Extraction

**Abstract:**

Saponins are naturally occurring, structurally diverse, surface-active glycosides produced by plants, marine animals and bacteria with many traditional and industrial applications in the food, pharmaceutical and healthcare industries because of the compoundsâ€™ naturally foaming and emulsifying properties. Grape pomace is a common solid by-product of the wine production industry. The objective of this work is to fill in the gap in present day literature on the properties of saponins found in grape pomace, specifically the extraction process, yield, and micellar properties from MourvÃ¨dre grape pomace. The first part of this study verified the existence of saponins in grape pomace qualitatively using the Liebermann Burchardt color change test and thin layer chromatography. Next, the saponins were extracted from ground grape pomace with organic solvents to yield 19.7% and 17.8% of grape saponins using batch and continuous extraction processes, respectively. Dye solubilization tests confirmed that the extracted product from grape pomace exhibits behavior of a surfactant due to the formation of micelles at the critical micelle concentration of 0.8 g/L. Lastly, the mass-transfer coefficient for both a single particle and for the laboratory set up for continuous extraction was calculated to be $1.68'10^{5}$ and $1.44'10^{-5}$ m/s, respectively. A theoretical mass transfer model for extraction was developed and computationally solved using COMSOL Multiphysics, in order to further study the extraction behavior of saponins from spherical grape particles and complex multiparticle systems.

# 538969 An Experimental and Computational Study of Saponin Extraction from Wine Grape Pomace

**Abstract:**

Saponins are naturally occurring, structurally diverse, surface-active glycosides produced by plants, marine animals and bacteria with many traditional and industrial applications in the food, pharmaceutical and healthcare industries because of the compoundsâ€™ naturally foaming and emulsifying properties. Grape pomace is a common solid by-product of the wine production industry. The objective of this work is to fill in the gap in present day literature on the properties of saponins found in grape pomace, specifically the extraction process, yield, and micellar properties from MourvÃ¨dre grape pomace. The first part of this study verified the existence of saponins in grape pomace qualitatively using the Liebermann Burchardt color change test and thin layer chromatography. Next, the saponins were extracted from ground grape pomace with organic solvents to yield 19.7% and 17.8% of grape saponins using batch and continuous extraction processes, respectively. Dye solubilization tests confirmed that the extracted product from grape pomace exhibits behavior of a surfactant due to the formation of micelles at the critical micelle concentration of 0.8 g/L. Lastly, the mass-transfer coefficient for both a single particle and for the laboratory set up for continuous extraction was calculated to be $1.68'10^{5}$ and $1.44'10^{-5}$ m/s, respectively. A theoretical mass transfer model for extraction was developed and computationally solved using COMSOL Multiphysics, in order to further study the extraction behavior of saponins from spherical grape particles and complex multiparticle systems.

# 532819 Comparison Data on Antioxidant Activities, Flavonoid and Mineral Content Analysis of *Artocarpus Altilis* Leaves at Different Maturity Stages

**Abstract:**

## Abstract

This study focused on the antioxidant activities and mineral content of *Artocarpus altilis* (breadfruit) leaves at different maturity stages, namely young leaves (YL), mature leaves (ML) and aging leaves (AL). DPPH radical scavenging activity and ABTS were done to measure the antioxidant activities. The TPC, TFC and mineral analysis were carried out to determine the changes of the compound at each stage of the leaves maturity. From the DPPH and ABTS assay, YL showed the highest antioxidant property followed by AL and ML. Meanwhile, the TPC and TFC were found to be the highest in YL, ML and AL. YL also showed the highest mineral contents (Zn, Se, Mn, Na, Fe, Mg, K) relative to both ML and AL. The absence of Hg and As, Cd, Pb, and Cr which were detected to be well below the detection limit of National Pharmaceutical Regulatory Agency (NPRA) for herbs, further promotes the potential of use of all three types of leaves as a nutritive ingredient for the food industry.

## Keywords

Bread fruit, antioxidant, flavonoid, mineral content, different maturity stages

# 536357 Influence of Chemical Structure of Compounds Present in Essential Oils on Their Antimicrobial Activity

**Abstract:**

Nowadays, development of healthier and safer alternatives in the food industry is an essential requirement during product and process design. The importance of essential oils (EOs) as food additives to control food spoilage and foodborne pathogenic bacteria has rapidly increased in recent years, since they constitute a natural source to develop new antimicrobial agents. Selected authors have reported synergistic, additive, or antagonistic antimicrobial effects when specific essential oils are mixed, where the antibacterial activity of the mixture depends on the EOs' composition and the target microbial strains. Moreover, some correlations between the molecular structure of several compounds of essential oils and their antimicrobial effect have also been reported.

The objective of this work was to evaluate the possible correlations between the antimicrobial effects of selected EOs with their physicochemical properties. A database of the antimicrobial effects of pure compounds present in EOs, EOs' composition and their minimum inhibitory concentrations against several bacteria (Gram positive or negative), was collected from selected reports. Some physicochemical properties, such as viscosity, vapor pressure, diffusivity in water, and partition coefficients, among others, were estimated by group contribution methods from the molecular structure of the pure components present in studied EOs. Every property estimation was developed by means of the ASPEN PLUS process simulator using the Joback-Reid and Marrero-Gani methods; furthermore, selected mixture properties were also estimated trough the UNIFAC method, taking into account the compounds present in every studied EO. Results were evaluated using the principal components analysis method, indicating the existence of relationships between the physicochemical properties and the EOs' antimicrobial effect on Gram positive or negative bacteria. The obtained results will contribute to the suitable design of food additives in the form of optimal antimicrobial agents from essential oils.

# Producing a Value Added Artificial Sweetener from Dairy Processing By-Product Via the Hydrogenation of Lactose

*Andrew Kasick [Presenter], Sunggyu Lee, Sustainable Energy and Advanced Materials (SEAM) Laboratory, Ohio University, Athens, OH*

## Introduction

### *Dairy whey waste and lactose*

In the dairy industry, cheese and yogurt manufacturing activities generate a liquid byproduct known as whey, which can represent between 80 to 90% of the volume of the milk originally used in the dairy processing[1]. Liquid cheese whey produced globally can amount to 145 million tons, and about 40% is not utilized industrially[1]. The composition of whey plays a significant role in determining its potential for industrial utility; sweet whey is typically the source of commercially available whey protein, while protein recovery from acid whey, the byproduct of yogurt production, is more difficult because of the higher acidity and mineral content, and lower available protein[2,3]. Appropriately handling and disposing of whey can be challenging. Whey which is not industrially utilized is often either loaded into an anaerobic digester, spread over farmland as a fertilizer, or added as a supplement to animal feed, but the application of these methods needs to balance potential risks; for instance, excessive whey spreading onto farmland can result in soil fouling from solids accumulation, and whey runoff can result in oxygen depletion in waterways[4].

When viewed from an environmental perspective, whey's lactose component, which can represent 70% of the solids composition of whey, is particularly concerning, given that it has been observed that the chemical oxygen demand of whey, typically around 60,000 to 80,000 ppm, is only reduced by 10,000 ppm after protein recovery[1,5]. Since lactose is a major contributor to whey's environmentally challenging properties, valorizing lactose as a feedstock for the production of valuable compounds is a way to promote more environmentally-conscious practices.

### *Potential of lactitol*

The hydrogenation of sugars involves the reduction of the sugar's carbonyl group to produce sugar alcohols, and the reaction can be performed catalytically in the presence of hydrogen gas; in the case of lactose, which is a disaccharide composed of galactose and glucose, the hydrogenation of the glucose moiety produces lactitol[6].

Lactitol has a number of properties which make it a viable candidate to serve as a reduced-calorie bulk sweetener. Relative to sucrose, lactitol is about 40% as sweet, meaning that for sucrose substitution lactitol can be pared with a more intense sweetener, such as aspartame or acesulfame K, to create a mixture that resembles sucrose's sweetness but lacks its calorie content[7]. Lactitol has comparable solubility to that of sucrose when dissolved in water, which is significant as solubility is an important parameter in manufacturing sweets[7]. Hygroscopicity is a property describing the water absorbed by a given compound, and for certain food products less moisture absorption results in a longer retention of freshness; because lactitol is notably less hygroscopic than sucrose, it can be an attractive ingredient for baked goods[7]. Compared to sucrose, lactitol elicits a significantly milder relative glycemic response (5 compared to sucrose's 68) and relative insulin response (4 compared to sucrose's 45), so lactitol it is regarded as a safe sugar alternative for a diabetic diet[8]. Lactitol is considered non-cariogenic, meaning that it does not contribute to tooth decay because it is not converted to dental plaque by bacteria in the mouth, another attractive property for a food additive[7]. In the U.S. lactitol has the status of being self-affirmed generally recognized as safe, and in 1983 the Joint FAO/WHO Expert

Committee of Food Additives (JECFA) approved lactitol's classification as a food additive as a sweetener, texturizer, and thickener[7,9].

Given the potential marketability of lactitol as an alternative to sucrose, and that the feedstock lactose could be derived from dairy whey byproduct, it is warranted to investigate the current state of knowledge regarding the hydrogenation of lactose to lactitol, and to research opportunities for process improvement.

***Overview of research***

Recognizing the motivating factors for both the increased utilization of lactose and the potential value of lactitol as a product, this research seeks to investigate previously unexplored aspects of the direct hydrogenation of lactose to lactitol, specifically what effect alterations to the solvent medium in which the reaction occurs have upon the rate of hydrogenation.

## Body

***Previous work from literature***

Previous studies of the hydrogenation of lactose to lactitol have primarily focused upon deriving kinetic expressions and evaluating how changing conditions affect the conversion[10–12]. Early studies have employed conventional hydrogenation catalysts, while later studies used the hydrogenation of lactose as an opportunity to evaluate the capabilities of novel developments in hydrogenation catalysts, such as new supporting materials and different metal impregnation techniques[13,14]. But modification to the catalyst is not the only route which can be taken to influence the occurrence of the hydrogenation reaction.

In the earlier studies which have been reviewed for lactose hydrogenation, the reactions were consistently performed in pure water. Concerning other sugar compounds, such as glucose and xylose, other authors have reported that modifications to the solvent medium in which the reaction occurs, specifically the addition of quantities of alcohol, such as ethanol or isopropanol, were capable of improving the hydrogenation rate of the studied sugar[15,16]. Explanations for how the selection of solvent affects the hydrogenation process varied. A popular reasoning was that the hydrogenation rate is improved in certain solvents over others because hydrogen is more soluble in certain solvents and less soluble in others, the implication being that a solvent with higher hydrogen solubility will increase the transfer of hydrogen to the catalyst surface, thereby improving the rate of reaction[16,17]. Later, alternative explanations were proposed, such as that different solvents interact to different extent with the reactant molecules, so that low interacting solvent-reactant combinations make the reacting species more available to adsorb onto the catalyst[18]. Other explanations contended that the interaction between the solvent molecules and the catalyst surface may contribute to differences in hydrogenation rate, as a solvent which adsorbs strongly to the catalyst surface could reduce the active sites available to promote the desired reaction[18,19].

***Preliminary work***

Since lactose does not demonstrate ready solubility in a wide range of common solvents, it was necessary to assess which solvents at what concentrations would be the most advantageous for studying alternatives to water. Since previous work has noted the viability for performing hydrogenation of other sugar compounds in aqueous alcohol solutions, assessing lactose's behavior in alcohols was a natural first step[15,16]. Lactose is significantly less soluble in ethanol than water, and it has been previously observed that as the ethanol concentration increases the amount of lactose which can enter solution decreases[20]. It has also been reported that as temperatures increase, the amount of lactose which can enter solution before reaching saturation increases[20]. To design experiments to study the hydrogenation of lactose in solvents of differing composition, it was important to establish the proportional quantities of the species in solution, and to determine a procedure to prepare the needed solutions.

For aqueous samples at 34°C (307 K) saturated with lactose, ethanol additions up to 20%v/v, or 16.5wt% ethanol, have not been observed to cause lactose crystal nucleation[21]. This would suggest that lactose and ethanol concentrations below these extremes would form a stable solution in which reactions could be performed. Based on reported experimental and modelling results, it was also predicted that for a 80:20 wt% water to ethanol solution on a sugar-free basis, 10wt% lactose could be present without insolubility issues arising[20]. To evaluate the veracity of these predictions, solutions of various lactose, water, and ethanol concentrations were prepared. Drawing procedurally from previously described work, it was determined that the best way to prepare the solution was to warm the water quantity to around 40 to 50°C and stir with an automatic mixer, add the lactose to the warm and mixing water, and, after the lactose has dissolved, slowly add the ethanol quantity[16]. The solutions were then allowed to sit with mixing at temperature for as long as five hours. Based on these tests, 10wt% lactose in 80:20 water:ethanol did not display lactose precipitation over the time the solution sat warm and mixed, while 20wt% lactose in 80:20 water:ethanol did display an extremely fine, glittery, precipitate, which was taken to mean that lactose crystals had formed. These observations confirmed the earlier predictions made while reviewing literature, established a procedure for preparing reactant solutions, and informed the design of later hydrogenation experiments. Having studied the behavior of lactose in mixtures of water and ethanol, it was important to consider how hydrogen gas would interact with mixtures of water and ethanol.

Previous authors have noted that hydrogen gas is more soluble in ethanol than in water, and that mixtures of ethanol and water offer still higher hydrogen solubility than water alone[16,19]. The available Henry's law constants demonstrated that a large concentration of hydrogen can dissolve in pure ethanol than in pure water[22]. For the purposes of this investigation it was important to predict the solubility of hydrogen in solutions as the ethanol concentration increases, and models are available to use which allow the Henry's law constant of binary mixtures to be predicted based upon the Henry's law constants for the pure components, the molar volumes of the pure components, and the real molar volume of the two-component mixture[23]. After determining the non-ideal molar volumes for a series of water and ethanol mixtures, their predicted Henry's law constants were determined, and the resulting hydrogen mole fraction could be determined. This hydrogen concentration information along with the lactose saturation limits reported in the literature were both plotted against the molar fraction ethanol in the given water-ethanol solution, and are presented in Figure 1 below.

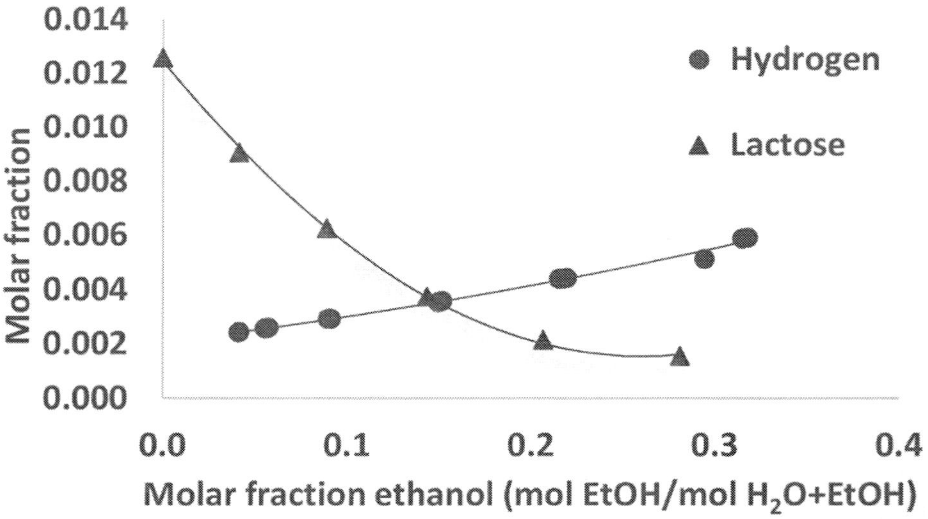


**Figure 1.** Impact of molar fraction ethanol upon molar fractions dissolved hydrogen and lactose.

This preliminary work demonstrated that as the molar fraction of ethanol in the solution increases, the concentration of dissolved hydrogen should be expected to increase while the limit of lactose solubility decreases. Based on prior work, it is expected that, for identical loadings of lactose, the hydrogenation rate should increase as the ethanol molar fraction increases.

***Future work for project***

In order to evaluate the above expectations of the hydrogenation of lactose, it is necessary to perform lactose hydrogenation experiments in solvent mediums of both water and mixtures of water and ethanol, and to characterize the product solution so that the amount of lactose converted to lactitol can be measured. Work is underway as the SEAM Lab to construct and operate a batch-wise hydrogenation experimental process system, equipped with a 300 c.c. autoclave reactor, to perform the prescribed experiments. Product mixtures are to be characterized using HPLC, operated with an appropriate column and liquid phase to achieve lactose and lactitol separation in a clear manner. By knowing the initial experimental conditions and the duration over which the hydrogenation was performed, it will be possible to evaluate the rate of the reaction, and by comparing similarly-condition experiments in pure water and a water and ethanol mix, any shift in hydrogenation rate should be detected.

## Conclusion

Whey, a liquid byproduct from certain dairy manufacturing activities, is a waste product which calls for appropriate handling and processing strategies. Lactose is whey's most abundant solid component, and the compound which significantly contributes to the byproduct's environmentally harmful properties, so researching and developing processing strategies which utilize lactose as a feedstock and produce value-added products is warranted. Lactitol, the sugar alcohol derivative of lactose, forms from the hydrogenation of lactose, and has a number of properties which make it an attractive alternative to the use of sucrose in food products. In the research which has previously studied the hydrogenation of lactose, the focus has been upon the evaluations of kinetics, the effect of varying processing conditions, or the application of novel developments in catalysts. Ongoing work at the Sustainable Energy and Advanced Materials (SEAM) Lab is taking a different approach, by considering what effect alterations to the solvent medium have upon the reaction; previous authors have observed that the addition of alcohols, such as ethanol and isopropanol to the reactant solution can improve the hydrogenation rate, but thus far this appears to be an unexplored topic in lactose hydrogenation. Preliminary work has characterized the predicted solubility behavior of lactose and hydrogen in aqueous solutions of increasing ethanol concentration, and future work will perform lactose hydrogenation experiments under the prescribed conditions and use HPLC to characterize the final products, in an effort to compare reaction rates.

## Acknowledgments

Much appreciation is owed to the Korea Institute of Science and Technology (KIST), the sponsors for this particular project, and Samwha Solutions, additional sponsors who support the operation of the Sustainable Energy and Advanced Materials (SEAM) Laboratory. Many thanks are owed to Dr. Sunggyu Lee, the Principal Investigator of the Laboratory, whose insight inspires and drives forward this hydrogenation project. Thanks are extended to Dr. Damilola Daramola, for his tireless efforts in managing the day to day operations of the Lab. Finally, much gratitude is given to the fellow members of the SEAM Lab, who always extended help and support when needed: graduate students Bahareh Baheri, Matthew Conley, Amy Lindenberger, and Lujie Ye and undergraduate students Michael Cole, Morgan Schroeder, Haley Vega, and Dillon Wiesler.

## References


1. Macwan, S. R.; Dabhi, B. K.; Parmar, S. C.; Aparnathi, K. D. Whey and Its Utilization. *Int. J. Curr. Microbiol. Appl. Sci.* [Online] **2016**, *5* (8), 134–155 DOI: 10.20546/ijcmas.2016.508.016 (accessed August 13, 2017).
2. Elliott, J. Whey Too Much: Greek Yogurt's Dark Side. *Modern Farmer* [Online] May 22, 2013 http://modernfarmer.com/2013/05/whey-too-much-greek-yogurts-dark-side/ (accessed August 2, 2017).
3. Smith, K. Wisconsin Center for Dairy Research. Current and Future Processing of Whey Ingredients [Online], 2015 http://oregondairy.org/wp-content/uploads/2014/04/4-2014-ODI.pdf (accessed August 2, 2017).
4. Prazeres, A. R.; Carvalho, F.; Rivas, J. Cheese Whey Management: A Review. *J. Environ. Manage.* [Online] **2012**, *110*, 48–68 DOI: 10.1016/j.jenvman.2012.05.018 (accessed October 24, 2017).
5. Gonzáles Siso, M. I. The Biotechnological Utilization of Cheese Whey: A Review. *Bioresour. Technol.* [Online] **1996**, *57* (1), 1–11 DOI: 10.1016/0960-8524(96)00036-3 (accessed October 30, 2017).
6. Murzin, D. Y.; Duque, A.; Arve, K.; Sifontes, V.; Aho, A.; Eränen, K.; Salmi, T. Catalytic Hydrogenation of Sugars. In *Biomass Sugars for Non-Fuel Applications*; The Royal Society of Chemistry: Croydon, UK, 2016; pp 89–133.
7. Zacharis, C. Lactitol. In *Sweeteners and Sugar Alternatives in Food Technology* [Online]; O'Donnell, K., Kearsley, M. W., Eds.; Wiley-Blackwell: New Delhi, 2012; pp 275–294. DOI: 10.1002/9781118373941 (accessed August 3, 2017).
8. Livesey, G. Glycaemic Response and Toleration. In *Sweeteners and Sugar Alternatives in Food Technology* [Online]; O'Donnell, K., Kearsley, M. W., Eds.; Wiley-Blackwell: New Delhi, 2012; pp 3–26. DOI: 10.1002/9781118373941 (accessed August 3, 2017).
9. LACTITOL. *Evaluations of the Joint FAO/WHO Expert Committee on Food Additives (JECFA)* [Online] http://apps.who.int/food-additives-contaminants-jecfa-database/PrintPreview.aspx?chemID=2877 (accessed Oct 25, 2018).
10. Hu, M.; Kurth, M. J.; Hsieh, Y.-L.; Krochta, J. M. HPLC and NMR Study of the Reduction of Sweet Whey Permeate. *J. Agric. Food Chem.* [Online] **1996**, *44* (12), 3757–3762 DOI: 10.1021/jf960321n (accessed May 16, 2016).
11. Kuusisto, J.; Mikkola, J.-P.; Sparv, M.; Wärnå, J.; Heikkilä, H.; Perälä, R.; Väyrynen, J.; Salmi, T. Hydrogenation of Lactose over Sponge Nickel CatalystsKinetics and Modeling. *Ind. Eng. Chem. Res.* [Online] **2006**, *45* (17), 5900–5910 DOI: 10.1021/ie0601899 (accessed July, 21, 2016).
12. Kuusisto, J.; Tokarev, A. V.; Murzina, E. V.; Roslund, M. U.; Mikkola, J.-P.; Murzin, D. Y.; Salmi, T. From Renewable Raw Materials to High Value-Added Fine Chemicals—Catalytic Hydrogenation and Oxidation of d-Lactose. *Catal. Today* [Online] **2007**, *121* (1–2), 92–99 DOI: 10.1016/j.cattod.2006.11.020 (accessed August 3, 2016).
13. Doluda, V. Y.; Wärnå, J.; Aho, A.; Bykov, A. V.; Sidorov, A. I.; Sulman, E. M.; Bronstein, L. M.; Salmi, T.; Murzin, D. Y. Kinetics of Lactose Hydrogenation over Ruthenium Nanoparticles in Hypercrosslinked Polystyrene. *Ind. Eng. Chem. Res.* [Online] **2013**, *52* (39), 14066–14080 DOI: 10.1021/ie401778y (accessed May 9, 2018).
14. Meyer, N.; Devillers, M.; Hermans, S. Boron Nitride Supported Pd Catalysts for the Hydrogenation of Lactose. *Catal. Today* [Online] **2015**, *241*, 200–207 DOI: 10.1016/j.cattod.2013.11.067 (accessed July 21, 2016).

15. Bizhanov, F. B.; Drozdova, R. B. Glucose Hydrogenation in Aqueous Isopropanol Solutions. *React. Kinet. Catal. Lett.* [Online] **1982**, *19* (1–2), 31–34 DOI: 10.1007/BF02065053 (accessed July 4, 2018).
16. Mikkola, J.-P.; Salmi, T.; Sjöholm, R. Effects of Solvent Polarity on the Hydrogenation of Xylose. *J. Chem. Technol. Biotechnol.* [Online] **2001**, *76* (1), 90–100 DOI: 10.1002/1097-4660(200101)76:1<90::AID-JCTB348>3.0.CO;2-E (accessed October 3, 2016).
17. Maxted, E. B.; Stone, V. 90. The Solvent Factor in Catalytic Hydrogenation. *J. Chem. Soc. Resumed* [Online] **1938**, 454 DOI: 10.1039/jr9380000454 (accessed May 12, 2018).
18. Rajadhyaksha, R. A.; Karwa, S. L. Solvent Effects in Catalytic Hydrogenation. *Chem. Eng. Sci.* [Online] **1986**, *41* (7), 1765–1770 DOI: 10.1016/0009-2509(86)87055-5 (accessed May 12, 2018).
19. Nikoshvili, L.; Shimanskaya, E.; Bykov, A.; Yuranov, I.; Kiwi-Minsker, L.; Sulman, E. Selective Hydrogenation of 2-Methyl-3-Butyn-2-Ol over Pd-Nanoparticles Stabilized in Hypercrosslinked Polystyrene: Solvent Effect. *Catal. Today* [Online] **2015**, *241*, 179–188 DOI: 10.1016/j.cattod.2014.01.045 (accessed May 12, 2018).
20. Montañés, F.; Olano, A.; Ibáñez, E.; Fornari, T. Modeling Solubilities of Sugars in Alcohols Based on Original Experimental Data. *AIChE J.* [Online] **2007**, *53* (9), 2411–2418 DOI: 10.1002/aic.11258 (accessed May 9, 2018).
21. Parimaladevi, P.; Srinivasan, K. Anti-Solvent Crystallization of Lactose Single Crystals from Alcohol-Aqueous Solution at Different Concentrations. *Int. J. ChemTech Res.* [Online] **2014**, *6* (3), 1595–1597 http://sphinxsai.com/2014/vol6pt3/2/(1595-1597)ICMCT14.pdf (accessed May 9, 2018).
22. Ghavre, M.; Morrissey, S.; Gathergoo, N. Hydrogenation in Ionic Liquids. In *Ionic Liquids: Applications and Perspectives* [Online]; Kokorin, A., Ed.; InTech: Rijeka, Croatia, 2011; pp 331–392. http://www.intechopen.com/books/ionic-liquids-applications-and-perspectives/hydrogenation-in-ionic-liquids (accessed October 16, 2018).
23. Shulgin, I.; Ruckenstein, E. Henry's Constant in Mixed Solvents from Binary Data. *Ind. Eng. Chem. Res.* [Online] **2002**, *41* (6), 1689–1694 DOI: 10.1021/ie010911x (accessed October 21, 2018).

# 529955 Physicochemical Properties, Macro- and Microanalytes Analysis of Gluten-Free Flour As Potential Functional Food Ingredients

**Abstract:**

## Abstract

The market of gluten free is growing as consumer nowadays choosing a gluten-free diet as personal preference. The suitability of the gluten free flour for use as food or food ingredients is highly depend on its physicochemical properties. In this study, the breadfruit, pineapple and sweet potato pulp were dried and ground into powder. The physicochemical properties such as bulk density, pH, water absorption capacity, oil absorption and gelling concentration of these fruit were studied. Mineral analytes of the flour were also determined. Results for bulk density from the least to the higher amount was 0.44 ± 0.03, 2.00 ± 0.12 and 2.4 ± 0.41 for breadfruit flour, sweet potato flour and pineapple flour, respectively. Pineapple flour showed the lowest pH value (3.9 ± 0.02), followed by breadfruit flour (6.3 ± 0.04) then sweet potato flour (6.3 ± 0.00). Sweet potato flour showed the highest value in oil absorption 3.03 ± 0.3) while showed the lowest value in water absorption capacity (3.75 ± 0.24). On the other hand, pineapple showed the lowest value in oil absorption (1.47 ± 0.03) while breadfruit flour showed the highest value in water absorption capacity (8.80 ± 0.61). Breadfruit flour showed gelling activity at 13% concentration compared to sweet potato flour and pineapple flour. The extinguish characteristic properties posses by these flour enable to be incorporated as functional ingredients.

Keywords: gluten free flour, mineral content, physicochemical properties

## 518862 Isosteviol: Synthesis through Typical Lewis Acid-Catalysis ($Fe^{3+}$) and Preparation Thereof Inclusion Complex with Γ-CD

**Abstract:**

Diterpenoid isosteviol (Ist) and its derivatives have gained a growing interest for broad therapeutic benefits, especially in recent years. In this paper, a preliminary study on preparation of isosteviol with typical lewis acid and its inclusion complex, namely γ-cyclodextrin/isosteviol (γ-CD/Ist), were evaluated. Firstly, Ist was prepared from stevioside (Stv) with Lewis acid. $Fe^{3+}$ was the optimal catalyst among five typical Lewis acid and the highest conversion of Stv of 100% and yeild of Ist of 83.2% were obtained based on the results of single factor experiments. In addition, in order to overcome Ist poor aqueous solubility, γ-CD/Ist was prepared by a co-evaporation method. The possible binding mode of Ist with α-CD, β-CD and γ-CD was identified using AutoDock software, respectively. Taking account of host molecular screening assay, γ-CD is the optimal host molecular. The solubility of Ist increased 185 folds at least, and formation of 1:1 γ-CD/Ist inclusion complex. HPLC, IR, TGA, XRD, SEM and NMR were evaluated for the γ-CD/Ist characterization. At ambient temperature, γ-CD/Ist complex aqueous solution is extremely stability with neutral and slight pH values after 30 days. To our knowledge, this is the first to prepare Ist using Lewis acid and improve thereof solubility through inclusion complex.

# 536225 Raspberry-Derived Treatment of Inflammatory Bowel Disease

**Abstract:**

Inflammatory bowel disease (IBD) affects over 1.6 million people in the United States, with patients suffering from persistent diarrhea, rectal bleeding, weight loss, and abdominal pain.[1,2] In healthy people, the intestinal epithelium provides a size- and charge-selective barrier that allows for the absorption of nutrients while blocking the transport of bacteria and pathogens.[3] Patients with IBD experience chronic inflammation of the intestinal epithelium and increased permeability of the intestinal barrier.[4] This is problematic, because as intestinal barrier function decreases, molecular selectivity is lost and toxins may permeate across the intestineâ€"potentially entering systemic circulation and further stimulating inflammation. Current treatments are limited to the management of IBD, and include antibiotics, steroids, and immunosuppressants, all of which can lead to long-term health problems with prolonged use.[5]

We have developed a raspberry-derived treatment for IBD that fortifies intestinal barrier function, treating one the sources of the diseaseâ€"a leaky gut. Raspberry extract was produced by freeze drying fresh raspberries, extracting them with ethanol/water followed by acetone, and lyophilizing the aqueous phase to produce a stable extract. The raspberry extract was found to reduce the transport of a small molecule permeation marker across the Caco-2 monolayer intestinal model, indicating its potential as a treatment for IBD. Using healthy mice as a preliminary *in vivo* model, rectally administered raspberry extract decreased intestinal permeability more than twofold and took effect in as little as two hours. To examine the raspberry-derived therapeutic in a diseased animal model, dextran sodium sulfate was used to induce colitis (a form of IBD) in mice. Rectally administered raspberry extract reduced the intestinal permeability of diseased mice more than twofold compared to diseased mice receiving only PBS. Importantly, raspberry extract is not associated with the side effects of convention IBD treatments because the body is already well equipped to handle the metabolites in raspberries. These results demonstrate the potential of a naturally derived raspberry therapeutic to treat IBD by decreasing the permeability of the intestinal barrier, a novel approach that addresses the source of the disease, a leaky gut, rather than treating symptoms.

1. Kaplan, G. G. (2015). The global burden of IBD: from 2015 to 2025. *Nature Reviews Gastroenterology and Hepatology*, 720-727.
2. Strober, W., Fuss, I., & Mannon, P. (2007). The fundemental basis of inflammatory bowel disease. *The Journal of Clinical Investigation*, 514-521.
3. Shen, L., Weber, C. R., Raleigh, D. R., Yu, D., & Turner, J. R. (2011). Tight Junction Pore and Leak Pathways: A Dynamic Duo. *Annual Review of Physiology*, 283-309.
4. Michielan, A., & D'Inca, R. (2015). Intestinal Permeability in Inflammatory Bowel Disease: Pathogenesis, Clinical Evaluation, and Therapy of Leaky Gut. *Mediators of Inflammation*, 1-10.
5. Moura, F. A., de Andrade, K. Q., dos Santos, J. C., Araujo, O. R., & Goulart, M. O. (2015). Antioxidant Therapy for treatment of inflammatory bowel disease: Does it work? *Redox Biology*, 617-639.

# 539928 Improved Performance of Biomimetic Membrane Integrated with the Aquaporins Modified with *in-Vitro* Genetic Incorporation of *P*-propargyloxyphenylalanine

**Abstract:**

Aquaporins (AQPS) are widely applied in biomimetic membranes for water recycling and desalination. To improve the stability of AQPS in biomimetic membranes, modified AQP Z and AQP SS9 were obtained by incorporation of an unnatural amino acid (*p*-propargyloxyphenylalanine, *p*Ppa) into AQPS *in vitro* using a mutated *Methanocaldococcus jannaschii* tyrosyl-tRNA synthetase (TyrRS) via the cell free expression system. The modified AQPS can realize the covalent-link with phospholipids in membrane and hence improve the stability of biomimetic membrane greatly. The concentration of $Mg^{2+}$ and fusion expression with signal peptides were evaluated to enhance the expression level of modified AQP SS9 and AQP Z. The modified AQP SS9 and AQP Z were then reconstituted into DOPC liposome and analyzed by stopped-flow spectrophotometer. The obtained water permeability coefficient ($P_f$) of $7.46x10^{-4}$ m/s was 5.7 times higher than that of proteoliposomes with the original AQP SS9 (Pf=$1.31x10^{-4}$ m/s), and 6.8 times higher than that of the DOPC liposomes (Pf=$5.01x10^{-5}$m/s). The membrane embedded with modified AQP SS9 has not only showed improved stability, but also exhibited a higher water flux and NaCl rejection, than those of natural AQP SS9. Similar results were also obtained with modified AQP Z. The research was helpful to develop biomimetic membranes with highly stable aquaporins.

# 534537 Phospholipid Bilayer Functionalized Membrane for Immobilized Enzymatic Catalysis

**Abstract:**

Enzymes, immobilized within membrane pores, offer improved kinetics due to pressure driven convective flow. On the other hand, phospholipid bilayer (PLB), owing to its fluidity and native â€œcell-membraneâ€ environment, facilitates the activity and stability of enzymes. In this study, we integrate the advantages of biomimetic characteristics of PLB and convective flow through microporous membrane within the same configuration (1). We report development of a novel bioinspired membrane by incorporating PLB within membrane pores using two different approaches â€“ (i) direct deposition of lipid bilayer within bare membrane pores, and (ii) immobilization of a polymer cushion prior to lipid bilayer within membrane pores. Developed PLB functionalized membrane pores were examined for enzymatic catalysis by electrostatically immobilizing glucose oxidase (GOx) enzyme. In-depth activity study revealed that the bioinspired membrane containing polymer cushion supported PLB and operating under convective mode of flow exhibited benefits in terms of activity, stability and operating range of pH. We also report the reusability of the membrane matrix as well as the PLB by exploiting the reversible attachment-detachment sequence of the functional moieties. Characterization of the detached phospholipid confirmed retention of the original structural and functional properties as exhibited before attachment. To the best of our knowledge, this is the first study that investigates the advantages of incorporation of PLB within membrane pores for immobilized enzymatic catalysis.

Reference: 1. A. Kumari, **S. Datta**[*], Phospholipid bilayer functionalized membrane pores for enhanced efficiency of immobilized glucose oxidase enzyme, *J. Mem. Sci.*, **539** (2017) 43-51

# 536956 Concentration of Polyphenols from Blueberry Pomace Extract Using Nanofiltration


**Abstract:**

Polyphenols extracted from blueberry (Vaccinium corymbosum) pomace were concentrated using nanofiltration. Crossflow filtration was shown to be a feasible method for concentrating the polyphenols present in dilute aqueous solutions. High-performance liquid chromatography was employed for the determination of total anthocyanins, total flavonols and chlorogenic acid in the hot water extract. Both nanofiltration membranes (NF245 and NF270) showed complete rejection of phenolic compounds at good permeances, whereas crossflow mode of filtration was found to reduce membrane fouling considerably. Furthermore, a suitable protocol was developed for clean-in-place of the used membranes. After repeated filtrations followed by the cleaning protocol, the rejection performance was preserved unaltered and the relative permeance was recovered up to 73% for NF245 membrane and more than 99% for NF270 membrane.


For commercial purposes, an ample amount of the blueberries is processed either into juice or juice concentrate and then the remaining solid residue (pomace) is generally treated as a waste product. The skins of blueberry fruits contain most of anthocyanins by weight so that, depending on the complexity of the juice extraction method, the pomace is left with a substantial amount of valuable polyphenolics. Therefore, there is an incentive to further process the blueberry pomace and extract those remaining polyphenolics for applications as natural colorants, encapsulated supplements or added nutraceuticals.

Most of the pomace extraction methods lead to dilute aqueous juice fractions and processing them into concentrates facilitates storage and transportation. Particularly, volume reduction and separation techniques are highly employed to produce juice concentrates and fractionate the dilute extracts. Multiple techniques have been developed aiming at the production of stable, nutrient-rich concentrated streams. Freeze concentration (cryoconcentration), osmotic distillation, membrane filtration and other multi-stage evaporation techniques are commonly used. Anthocyanins are labile compounds that have been shown to easily degrade and lose biological activity under severe processing parameters such as high temperatures, UV radiation and cross reaction with other processing chemicals. Therefore, a careful consideration has to be given to the choice of processing techniques which should not only be economically feasible but also limit the deactivation and loss of the bioactive compounds. Membrane technology could be a promising technology for recovery of these fragile biologically active compounds without destroying the aroma, colour, taste and other nutrient quality.

Frozen blueberry pomace was allowed to thaw to 21°C prior to extraction. A Dionex model 200 accelerated solvent extractor (ASE) system interfaced with a solvent controller (Dionex Corp., Sunnyvale, CA) was used to extract anthocyanins from blueberry pomace. Samples (0.5 g) were loaded into 22 mL stainless steel extraction cells with a cellulose paper filter inserted at the bottom of the cells. The ASE extraction was carried out using water as solvent: 68 bar, 120°C, five extraction cycles, 70% flush volume, 90 sec nitrogen purge time (no static time and no preheat time).

For each extraction cycle it took approximately 5-6 min for the water to heat to 120°C for a total run time of 25-30 min. Approximately 22 mL of extract from each extraction cycle was pooled after passing through a large microporous sieve. Pressurized hot water extracts were stored at -20°C prior to total anthocyanin analysis and nanofiltration testing. Non-prefiltered or prefiltered (0.22 μm and 0.45 μm) blueberry pomace hot water extracts are collectively described here as feed solution. The extract total anthocyanin concentration was found to naturally vary in the range 85-125 mg/L.

The HPLC data for total anthocyanins, flavonols and chlorogenic acid in the ASE extract, retentate and permeate revealed that the highest amount of polyphenols was due to total anthocyanins (with over 82% wt.), while the highest amount of sugars was due to fructose (with over 70% wt.) in the feed (ASE extract). The HPLC chromatograms of 16 anthocyanins and 7 flavonols were identified. The most prevalent ACYs were malvidin-3-galactoside (15% wt, 528.9 g/mol), delphinidin-3-galactoside (14% wt, 500.8 g/mol) and malvidin-3-glucoside (13% wt, 528.9 g/mol). The most prevalent FLA were quercetin-3-galactoside (37% wt, 464.4 g/mol), quercetin-3-glucoside (16% wt, 464.4 g/mol) and quercetin-3-acetyl rhamnoside (11% wt, 490.4 g/mol). Chlorogenic acid (354.3 g/mol) represented only 6% of total polyphenols. Both NF270 and NF245 exhibited complete retention of total polyphenols, and sucrose. Only small amounts of glucose and fructose were found in the permeate fractions so that the rejections were higher than 97%.

The nanofiltration membranes were tested at extended filtration times until the maximum amount of feed volume could be removed in a dead end mode. As permitted by the dead-end setup, a minimum of 20% v/v of initial feed should remain in the pressure vessel to not disrupt mixing. This analysis allowed to compare the performance of the two membranes in terms of permeance, rejection, anti-fouling properties, and also to investigate degradation of polyphenols at longer reactor residence time. NF270 showed better performance than NF245. NF270 required approximately 19 hours to remove 80% of the initial volume and the polyphenols content was concentrated by a factor of 4.6. In the same amount of time NF245 reduced the volume by 60% and concentrated the polyphenols by a factor of 2.2. It required almost 30 hours to reduce the feed volume to the same performance as with NF270, but the flux started to decrease considerably after 21 hours, due to increased fouling. The temperature of the feed and retentate was monitored at the start and at the end of each filtration and it was found to not change by more than ±1.0°C. Both membranes showed complete rejection of polyphenols and during analysis no polyphenol degradation was detected. However, both NF270 and NF245 showed adsorbed polyphenols on their surfaces, as observed from the dark purple color of the used membranes. This could be an effect of particle agglomeration and polyphenol adsorption as seen previously in unstirred systems.

Previously, the filtration performance in dead-end mode was tested under different experimental parameters and those findings were determinant in the design of the crossflow setup. The setup was constructed to keep the feed continuously stirred and the crossflow rate was set at a maximum flowrate of 57 mL/min. Then, the feed was passed through a 0.22 μm prefilter to remove larger aggregated particles. Total polyphenols were rejected completely and no polyphenol degradation was observed during analysis. The permeance was considerably higher with NF270, which started at 3.5 $L\ m^{-2}\ h^{-1}\ bar^{-1}$ and then reached approximately 2.0 $L\ m^{-2}\ h^{-1}\ bar^{-1}$ after 3 hours of filtration. For NF245 the permeance started at 1.5 $L\ m^{-2}\ h^{-1}\ bar^{-1}$ and then decreased to 0.6 $L\ m^{-2}\ h^{-1}\ bar^{-1}$ after the same filtration time.

For the purpose of membrane reconditioning, the used membranes were cleaned with the following protocol: (1) washed in deionized water for 24 hrs; (2) washed in 0.2% w/v HCl and (3) washed with 0.1% w/v NaOH. After steps (2) and (3) the membrane was cleaned-in-place with deionized water until the pH was constant. While establishing the wash protocol it was observed that step (2) leaves the previously purple stained membrane slightly pink in color, step (3) changes the staining to light brown and, for the membranes used in crossflow, the latter staining was eventually removed completely. The membranes used in dead-end maintained the light brown color and the fouling was irreversible. Only 37% of the initial water permeance could be recovered for NF270 and less than 20% for NF245.

# 539907 Efficient Production of (Z)-α-Santalol with Multi-Pathway Engineering in *Saccharomyces Cerevisiae*

**Abstract:**

Sandalwood oil fragrance is one of the world's most highly prized fragrances produced by tropical sandalwoods, which is not only broadly applied in the fragrance industry, but also has various biological activities, such as anti-cancer, anti-tumor, anti-viral, anti-diarrhoeal, etc. (Z)-α-santalol and (Z)-β-santalol are the most important contributors to sandalwood oil fragrance. Microbial cells engineered for efficient production of plant sesquiterpenes may allow for sustainable and scalable production of these compounds. As a multi-substrate P450, SaCYP736A167 stereo-selectively produces (Z)-a-santalol from α-santalene. In this study, we constructed the (Z)-α-santalol pathway by heterologous expressions of a truncated form of 3-hydroxyl-3-methyl-glutaryl-CoA reductase (HMGR), a santalene synthase (SaCYP736A167), and an NADPH-dependent cytochrome P450 reductase (CPR) in *Saccharomyces cerevisiae*. Further efforts were paid to increase the production of (Z)-α-santalol production with down regulation of the competitive pathways by expressing *ERG9* in the ergosterol pathway with the weak HXT1 promoter and deleting *LPP1* and *DPP1* in the farnesol pathway to divert the carbon flux from sterol synthesis towards a-santalene. Finally, with the fusion of eleven different signal peptides targeting mitochondrion with 8 genes in MVA pathway and santalene synthas to boost acetyl-CoA utilization and the deletion of *PAH1* to expand the nuclear/ER membrane to anchor more SaCYP736A167, the production of (Z)-α-santalol was improved greatly. The present work paved one road to produce this attracting fragrance component in industry.

# 529971 Functional Characterization of Soypeptides As Supplementary Diet and Their Effects on the Kinetics of Cell Growth of Probiotic Microorganisms

**Abstract:**

## Abstract

Demand from consumers on functional food products containing probiotic and prebiotic remarkably increasing as it confers health benefits and help to prevent chronic diseases. The desired probiotic product is usually made of a combination of high concentration of viable probiotic strains and a suitable prebiotic compounds which help in the growth and colonization of the probiotics after consumption. Nowadays, soypeptides are considered as one of the important in food and nutraceutical industries based on their high nutritional values as protein source. Therefore, soypeptides are widely used in many food formula or consumed separately as food supplement. The aim of this study is to investigate the functional characteristics and potential use of soypeptides as prebiotics. Thus, the effect inulin and soypeptides (when used separately or in combination) on the growth kinetics of different groups of probiotics was investigated. This study involved wide range of microorganisms which represent different classes of probiotics (*Lactobacillus sp., Bifidobacterium sp.* and probiotic yeasts). In addition, the antioxidant activity of soypeptides was investigated using ABTS and DPPH methods. The results showed that soy peptides positively affect the growth rate of different probiotic microorganisms when used separately or in combination with inulin. They also exhibited high antioxidant activities which considered as more added value in food formula. The results of this work provide strong evidence of the importantce of soypeptides in human diet and their high potential applications as main ingredient in functional food industries.

Keywords: soy peptides, probiotics, prebiotics, growth kinetics, antioxidant activities.

# 532085 N-Butanol Production from Cotton Stalk Using Engineered *Clostridium Cellulovorans*

**Abstract:**

Cotton stalk (CS) is a potential biomass for n-butanol production, but pretreatment and enzymatic hydrolysis are essential steps to obtain fermentable sugars because of the recalcitrant nature of lignocellulose, which usually results in a high operational cost. In this study, a consolidated bioprocess (CBP) was developed using a metabolically engineered *Clostridium cellulovorans* adhE2. First, various saccharides as substrates for n-butanol production by *Clostridium cellulovorans* adhE2, including glucose, xylose, cellulose, xylan, and xylooligosaccharide, were investigated. The results showed that n-butanol can be produced from all substrates studied, including cellulose and xylan.. However, faster growth appeared in media containing glucose and xylose. Second, various methods for the pretreatment of CS, i.e. dilute sulfuric acid pretreatment (DSAP), high pressure-assisted alkali pretreatment (HPAP), and high temperature water pretreatment (HTWP), and subsequent n-butanol fermentation from pretreated CS, without enzymatic hydrolysis, were explored. It was found that the intact structures of pretreated CS were obviously disrupted. The lignin of biomass was largely removed and the crystallinity of cellulose decreased after pretreatments. HPAP led to the highest reducing sugar and n-butanol yields compared to UAAP and HTWP. Finally, the engineered *C. cellulovorans* adhE2 produced n-butanol with a final concentration of >2 g/L and yield of >0.09 g/g from pretreated CS in CBP. This study demonstrated that the fermentation process with the engineered *C. cellulovorans* enabled a high n-butanol production directly from lignocellulosic biomass.

# 538756 Oxidation-Reduction Potential Controlled Microaeration for Fermentation of Lignocellulose Feedstock

**Abstract:**

The dilute acid pre-treatment of pine lignocellulose produces inhibitors (acetic acid, furfural, and 5-hydroxymethylfurfural) that decrease specific cell growth rate and ethanol productivity during fermentation. Concentration of this fermentation feedstock via nanofiltration results in increased initial glucose concentrations and higher final ethanol titers, both of which further inhibit fermentation. Previous work with very high gravity fermentation has shown that oxidation-reduction potential (ORP) controlled microaeration (MA) increases cell growth rate and ethanol productivity when the fermentation is inhibited by glucose and ethanol. The goal of this study was to apply this same ORP-controlled microaeration strategy to a mock lignocellulose hydrolysate feedstock (LHF) to determine the effect of microaeration on fermentation kinetics, and to determine if microaeration can reduce fermentation costs compared to traditional unaerated (UA) (strict anaerobic) fermentation. It was found that microaeration (ORP controlled at -150 mV) during LHF fermentation increased the specific ethanol productivity by 250% ($Q_{UA,P}$=0.51 vs. $Q_{MA,P}$=1.32) during exponential growth with no reduction in ethanol yield ($Y_{P/S}$=0.49). Additionally, higher specific inhibitor reduction rates (furfural; 0.13 vs. 0.25 g/L/OD/hr, and HMF; 0.02 vs. 0.06 g/L/OD/hr, for UA vs MA, respectively) were observed. When the experimental results were applied to an economic assessment for a 5 MGPY LHF to ethanol plant, it was found that microaeration resulted in an increase in fermentation EAOC of $2.19 million/yr. The results indicate that microaeration is not a more economical mode of operation for LHF fermentation than traditional unaerated fermentation.

# 539744 Effects of Artificial Electron Carriers on High-Efficient Butyric Acid Production through Co-Fermentation of Glucose and Acetate By *Clostridium Tyrobutyricum*

**Abstract:**

The addition of acetate into Clostridial Growth Medium during glucose fermentation (at 24 h) was found to inhibit acetate formation and increase the final titer and productivity of butyrate by *Clostridium tyrobutyricum*. However, only limited (0.79-7.62 g/L) acetate was utilized. In order to strengthen the acetate re-assimilation, artificial electron carrier (methyl viologen, benzyl viologen or neutral red) was also introduced into the medium to increase the NADH supply which facilitated the acetate utilization. As expected, more acetate was utilized (4.9-20.3 g/L) and enhanced butyric acid fermentation was achieved. Subsequently, the effect of acetate and glucose co-fermentation with benzyl viologen addition was studied in bioreactor. Compared with the control, when 13.4 g/L acetate and 3.75 Î¼M BV were added into the medium at 27 h, the final concentration, productivity and yield of butyrate increased by 40.5% (from 39.8 to 55.9 g/L), 10.0% (from 0.69 to 0.76 g/L/h) and 32.3% (from 0.31 to 0.41 g/g carbon source), receptivity. In addition, the selectivity of butyrate exhibited a sharp increase from 78% to 97.4%. Therefore, the proposed strategy can be used to obtain high-efficient butyric acid production. Furthermore, acetic acid in-suit detoxification which enables commercial utilization of lignocellulosic biomass hydrolyzates could be achieved by artificial electron carrier addition.

# 531995 System Metabolic Engineering ofÂ *Clostridium Cellulovorans* Towards Consolidated Bioprocessing for *N*-Butanol Production from Cellulosic Biomass

**Abstract:**

*n*-Butanol can be used as a significant industrial chemical and potential superior biofuel. Especially, compared with ethanol, *n*-butanol can be used as an ideal substitute of gasoline because of its high energy density, low water solubility, and low vapor pressure. Lignocellulosic biomass is an abundant, cheap, and renewable carbon source, and is a desirable feedstock for biofuels production. In our previous research, *Clostridium cellulovorans*, a natural cellulose/hemicelluloses utilizing, butyrate producing bacterium, was metabolically engineered for n-butanol production from cellulose by overexpressing *adhE2* from *Clostridium acetobutylicum*. Therefore, consolidated bioprocessing (CBP) with combined enzyme production, biomass hydrolysis, and sugar fermentation in one step can be achieved by using the engineered *C. cellulovorans*. However, this mutant only produced 1.42 g/L n-butanol and 1.60 g/L ethanol from microcrystalline cellulose in 10 days. In this study, genome-scale metabolic model with omics analysis was firstly applied for guiding rational metabolic engineering of *C. cellulovorans*. System metabolic and process engineering, including optimization of butanol biosynthesis pathway, increasing C4 carbon flux, weakening competitive acids synthesis pathway, and improving the intracellular NADH availability through redox rebalancing, were carried out to increase butanol production in *C. cellulovorans*. The mutant produced 5.2 g/L *n*-butanol with a high yield of >0.3 g/g cellulose in batch fermentation, which were the highest ever achieved. This study demonstrated that *C. cellulovorans* is a promising CBP host for biobutanol production from cellulosic biomass.

# 539739 Improving the Fermentation Performance of *Clostridium Acetobutylicum* ATCC 824 By Strengthening the VB1 Biosynthesis Pathway

**Abstract:**

Vitamin B1 (VB1) is an essential coenzyme for carbohydrate metabolism and involved in energy generation in most organisms. In this study, we found that insufficient biosynthesis of VB1 in *Clostridium acetobutylicum* ATCC 824 is a major limiting factor for efficient acetone-butanol-ethanol (ABE) fermentation. In order to improve the fermentation performance of *C. acetobutylicum* ATCC 824, the VB1 biosynthesis pathway was strengthened by overexpressing the *thiC*, *thiG* and *thiE* genes. The engineered strain 824(thiCGE) showed enhanced VB1 and energy synthesis, resulting in better growth, faster sugar consumption, higher solvents production and lower acids formation than the wild type strain in both VB1 free and normal P2 medium (1 mg/L). Compared with the wild type strain, 824(thiCGE) produced 18.4% or 16.7% more butanol in VB1 free P2 medium when glucose or xylose was used as the substrate, respectively. When mixed sugar (glucose:xylose=2:1) was used as the substrate in VB1 free P2 medium, the xylose consumption rate and butanol titer of 824(thiCGE) were 46.2% and 20.0% higher than those of the wild type strain. All these results demonstrated that this metabolic engineering strategy could provide a new and effective way to improve the cellular performance of solventogenic clostridia. In addition, it may have some potential application value in ABE fermentation using simple medium and/or lignocellulosic biomass.

# 533199 Optimization of Pleuran Production By Pleurotus Ostreatus Using Batch and Fed-Batch Cultivation System

**Abstract:**

*Pleurotus* mushrooms are known for producing Î²-glucans which have medicinal properties as a constituent in the cellular cell wall of the fruiting body or the mycelium. This mushroom is rich with protein, fibre, carbohydrates, vitamins and minerals. Several bioactive compounds have been identified in *P. Ostreatus* showing strong anti-cancer activities such as polysaccharides, mainly Î²-D-glucans, proteoglycans, lectins, fibres, terpenoids, steroids, nucleic acids and others. Pleuran is an exopolysaccharide (Î²-(1,3/1,6)-D-glucan) produce by *P. Ostreatus*. This exopolysaccharide has molecular formula of $(C_6H_{10}O_5)_n$ while the molecular weight is varied depending on the length of molecular structure repetition. The main objective of this research is to develop a new process for maximal production of pleuran by *P. Ostreatus* in submerged fermentation both in shake flask and bioreactor. In media screening, highest pleuran production of 2.13 g $L^{-1}$ obtained in selected medium which consist of (in g $L^{-1}$): glucose, 60.0; yeast extract, 3.0; $(NH_4)_2SO_4$, 2.0; $K_2HPO_4$, 1.0 and $MgSO_4.7H_2O$, 0.2. The medium component was then optimized using One-Factor-at-a-Time and statistical method. The optimized medium using OFAT which composed of (in g $L^{-1}$): glucose, 100.0; yeast extract, 3.0; $(NH_4)_2SO_4$, 2.0; $K_2HPO_4$, 1.0; $MgSO_4.7H_2O$, 0.2 was able to produced 3.33 g $L^{-1}$ plueran. The OFAT optimization was done to provide better understanding on suitable parameter range during statistical optimization. For semi industrial scale production, statistical optimized medium was used in 16-L bioreactor fermentation. The batch cultivation was conducted in order to understand the growth pattern of *P. Ostreatus* and to design the cultivation strategies for fed-batch cultivation. The batch cultivation of *P. Ostreatus* was studied under controlled pH condition. Following that, fed-batch cultivation was carried out using constant feeding strategy of 0.4 g $L^{-1}$ $h^{-1}$ substrate. The substrate tested were either by using mono-glucose or full medium composition. It was found that feeding with full medium composition provide better result compared to feeding with mono-glucose. Feeding using full medium produced 4.5 g $L^{-1}$ pleuran while feeding with mono-glucose produced only 3.8 g $L^{-1}$ pleuran. In conclusion, this study offers a promising and comprehensive strategy for development of new process for maximal pleuran production by *P. Osteratus* in submerged fermentation in both shake flask and scale-up in 16-L bioreactor.

# 537003 Extraction, Purification and Modification of Poly (3-hydroxybutyrate) Produced By the Fermentation of Fatty Acids with *Burkholderia Cepacia B27*

**Abstract:**

Synthetic polymers are composed by monomeric units that came from fossil hydrocarbons, this materials are widely used around the world due to its low production cost and good mechanical properties. The main application of this materials is packaging, which represents a short useful life in most cases. Usually, many packages are used one, or a couple times before they are discarded, this and the non-biodegradable behavior of most synthetic polymers used for packaging caused the amount of plastics in landfills to grow from 1% in 1960 to 10% in 2005, and the accumulation of millions of tons of plastic waste in water bodies all around the world [1]. That is why in recent years, the interest in the synthesis of materials able to replace the conventional polymers but with high biodegradation rates and low toxicity has grown.

Polyhydroxialkanoates (PHAs) are thermoplastic polyesters synthetized by a large group of bacteria under stress conditions (limitation of nutrients), to store energy as intracellular fat deposit; they are biodegradable and non-toxic, and some of them have interesting mechanical properties that allow them to be used in the packaging industry. Poly (3-hydroxibutyrate) or P3HB is a linear homochiral, and crystalline biopolymer that belongs to a group of short chain PHAs [2]. Itâ€™s widely studied due to its mechanical properties that are similar to those of polypropylene [3], but with a better oxygen barrier. It is also resistant to hydrolytic degradation, which is a feature that many other biopolymers like PLA or starch-based polymers lack. However, itâ€™s necessary to improve some of its weakest characteristics like the high brittleness, the production cost and the narrow processing window. This improvements can be achieved using different extraction, purification and modification technics that result in materials with different physicochemical characteristics and production costs [4].

In the Biotechnology institute of the National University of Colombia research works have been developed in the production of the biopolymer Poly (3-hydroxybutyrate) from strains isolated from Colombian soils. This strains were evaluated from a microbiological and operational point of view and the strain *Burkholderia cepacia 2G57* was selected and genetically modified to create the P3HB hyperproductive mutant *Burkholderia cepacia B27 [5]*, also, studies to optimize the production process were carried out, evaluating different operational conditions in 7L, 100L and 2000L fermentations [6] and different extraction and purification techniques like the digestion of the biomass with SDS and the purification of the polymer with $CHCl_3$ and NaClO obtaining as result a yellowing material with strong odor and apparent low molecular weight due to its high fragility and poor mechanical properties. In this work different purification processes were evaluated in order to obtain a high purity polymer. Initially, the polymer was extracted and separated from the surrounding PHA hyper-productive mutant bacteria *Burkholderia cepacia B27* biomass, using chemical digestion with SDS, centrifugation and solvent precipitation with ethanol. Then the polymer was purified to remove protein and oil residues from the fermentation, for this the performance of different solvents such as ethanol and methanol was tested under different operation conditions. High purity polymer with good mechanical properties, colorless and odorless was obtained. All the samples were thermally characterized (TGA and DSC) to evaluate purity and crystallinity of the resulting materials.

**References**

[1] R. Geyer, J. R. Jambeck y K. Lavender Law, «Production, use, and fate of all plastics ever made,» *Science Advances,* vol. 3, nº 7, 2017.

[2] A. P. Nayak, «Biodegradable polymers: opportunities and challenges,» *Journal of Macromolecular Science,* vol. 3, nº 39, pp. 481 - 505, 1999.

[3] D. Parra, J. Fusaro, F. Gaboardi y D. Rosa, «Influence of poly (ethylene glycol) on the thermal, mechanical, morphological, physicalechemical and biodegradation properties, of poly (3-hydroxybutyrate),» *Polymer DEgradation and Stability,* pp. 1954 - 1959, 2007.

[4] V. P. Cyras, A. Vasquez, C. Rozsa, N. Galego Fernandez, L. Torre y J. M. Kenny, «Thermal estability of P(HB-co-HV) and its blends with polyalcohols,» *Journal of Applied Polymer Science,* vol. 77, nº 13, pp. 2889-2900, 2000.

[5] D. A. Mendez, «Mathematical Modelling and Scale-up of Batch Fermentation with Burkholderia cepacia B27 Using Vegetal Oil as Carbon Source to Produce Polyhydroxyalkanoates,» *Chemical Engineering Transactions,* vol. 49, pp. 277-282, 2016.

[6] D. A. Mendez, I. O. Cabeza y N. C. Moreno-Sarmiento, «Volumetric Oxygen Mass Transfer Coefficient Determination and Hydrodynamic Optimization of Polyhydroxyalkanoate Production with Vegetal Oil as Carbon Source,» *Chemical Engineering Transactions,* vol. 57, pp. 1303-1308, 2017.

# 532112 Analysis and Design of Kinetic Controls of Fatty Acid Synthesis

**Abstract:**

Microbial systems offer a promising means of producing fuels and chemicals from renewable feedstocks. Enzymes that evolved to control cellular growth and maintenance can be repurposedâ€"or re-engineeredâ€"to convert metabolic intermediates into a wide range of useful oleochemicals (e.g., fatty acids, alkanes, alcohols, alkyl esters, and methyl ketones). To date, most approaches to synthesize these products have used genetic adjustments (i.e., enhanced or attenuated expression of native and non-native genes) to make additive or subtractive changes to the composition of the fatty acid pathwayâ€"often without consideration for the kinetics of the pathway. These approaches have enabled the production of useful classes of chemicals, but have proven limited in their ability to exert tight control over product profiles. In this talk, we will introduce a detailed kinetic model of type II fatty acid synthesis, and we will show how this model can facilitate metabolic engineering of fatty acid production in microbial systems. The results of our analysis explain unexpected observations from experimental studies of fatty acid synthesis (e.g., inhibition of fatty acid production by high concentrations of acyl carrier protein [ACP] and fatty acid biosynthesis [Fab] enzymes), provide an analytical framework for identifying kinetically coupled biocatalytic steps, and reveal new strategies for pairing coordinated genetic adjustments with protein engineering to tune fatty acid profiles in *Escherichia coli*.

# 538310 Rewiring *Yarrowia Lipolytica* Lipid Metabolism for the Production of Omega-3 Fatty Acid Using Alternative Substrates

**Abstract:**

Omega-3 long chain fatty acids are essential for human health and development. The stagnant production of omega-3 fatty acids from fish oils no longer meets the demands of the fast growing aquaculture and nutraceutical market. A sustainable and economic source of omega-3 fatty acids is urgently needed. Here we report a systematic approach to metabolic engineering of non-conventional yeast *Y. lipolytica*for eicosapentaenoic acid (EPA) production from non-sugar substrates such as cheap plant oils and rendered animal fats. We first established an alternative EPA biosynthesis pathway into *Y. lipolytica*by heterologous expression of a set of elongases and desaturases utilizing acyl-CoA instead of phospholipids, resulting in efficient production of EPA while avoiding accumulation of undesired intermediate fatty acids. We also enhanced the assimilation and synthesis pathways of triacylglycerides, and disrupted its degradation pathway to improve the efficiency of utilizing exogenous lipids as building block for EPA production. By redirecting carbon flux towards fatty acid elongation and desaturation, we further increased the amount of EPA that was incorporated into triacylglycerol. Finally, we have tested several strategies to increase acyl-CoA flux, including enhancing acyl editing and fatty acid transport. Along the way, novel genes in the Y. lipolytica genome were identified that enhanced acyl-CoA flux. Our results illustrate that the acyl-CoA pathway circumvents bottlenecks in the conventional phosphatidylcholine dependent desaturase pathways necessary for accumulation of triacylglycerides with high EPA content. This also suggests a new paradigm for the use of low-value alternative feedstocks for bioprocesses.

# 535043 Exploiting the L-Lactate Biosynthetic Pathway inÂ *Corynebacterium Glutamicum*Â for Heterologous Production of D-Lactate from Biomass-Derived Carbon Substrates

**Abstract:**

Polylactic acid (PLA), bio-based polymers gained attention as an economically feasible packaging material like fresh fruit container, drinking cups, lamination films, and many more. The skeletal formation of PLA involves the lactic acid as a primary precursor. In the present study, various heterologous strains of *Corynebacterium glutamicum (*AV41, AV42, AV43, AV44, AV46) were generated to produce D-lactate by exploiting the L-lactate production pathway and followed by ldhA overexpression in the biosynthetic pathway. The heterologous strains with the constitutive promoter were grown in different production media supplemented with different level of different carbon sources. The strain AV46 produced 5887 mg/ L of D-lactate from a minimal media supplemented with glucose as the carbon source. The strains were grown in biomass hydrolysate, and a similar level of D-lactate production 5276 mg/ L was observed. Moreover, upon optimization of pH, the D-lactate production from biomass hydrolysate was enhanced by 1.3 fold, 6945 mg / L at pH 8 in strain AV46. For further utilization of biomass hydrolysate and enhancement of D-lactate production, the strain AV 46 was cultured in different aromatics and found to be growing well without any sugar supplementation. In the preliminary results, we found the strain to be producing D-lactate from aromatics such as benzoic acid, cinnamic acid, vanillic acid, and coumaric acid directly. Additional heterologous proteins are to be overexpressed from xylose and arabinose pathway to increase the ability of these strains to utilize xylose and arabinose as carbon substrate. With the conversion of all major carbon substrates (glucose, xylose, arabinose, and aromatics) in the renewable biomass to D-lactate, we aim to increase the yield, titer, and productivity of D-lactate from biomass hydrolysates.

# 527502 Homologous Constitutive Expression of Halophilic and Acidophilic Î²-Glucosidases in Marine *Aspergillus Niger* Zjube-1

**Abstract:**

**Abstract:** Î²-glucosidase (BGL) is widely applied in biofuel industry, as a part of cellulase cocktail to catalyze the hydrolysis of the Î²-1,4 linkages that join two glucose molecules in a cellulose polymer. The hydrolysis step by BGL is generally recognized as the major limiting step in cellulose degradation and the BGLs with prominent enzymatic properties are of great importance for the efficient utilization of lignocellulosic biomass.

In this study, two *bgl* genes (*bgl1* and *bgl2*) were cloned from the genome of marine *Aspergillus niger* ZJUBE-1. Then two *bgl* expression cassettes driven by *gpdA* promoter were respectively transformed into marine *A. niger* for homologous constitutive expression. For few protein was secreted during fermentation with original strain at non-induced condition, the constitutive expressed BGLs (BGL1 and BGL2) were dominated in the fermantation broth. Conveniently, the purification of BGLs was accomplished by ultrafiltration and anion exchange chromatography and the purity was up to 95%. Subsequently, the study of enzymatic properties showed that BGL1 and BGL2 exhibited maximum activity at pH 3.0-4.0 and pH 3.5-4.5, respectively, suggesting these two BGLs were acidophilic, especially for BGL1. Notably, the activity of BGL2 increased by 44% in the presence of 4 M NaCl compared with the activity in the absence of NaCl, suggesting BGL2 was halophilic. Furthermore, the thermostability of BGL2 improved in the presence of NaCl. Incubated at 60 °C for 2 h, BGL2 in 4 M NaCl maintained 81 % activity compared with 22% activity in the absence of NaCl. At last, the key amino acid residues involved binding and catalysis were determined by homology modeling to these two BGLs. Hypothetically, the difference of asparagine residues in substrate-binding pocket changed the properties of hydrogen bonds between pocket and substrate, and then endowed the halophilism to BGL2.

Marine microorganisms are the treasury of enzymes with special properties, especially the halophilic enzymes. The halophilic BGL2 obtained from marine *A. niger* can be applied in some catalytic processes with high salt concentration and the structure study on BGL2 can provide the basis for other halophilic enzymes.

**Keywords:** Î²-glucosidase, marine *Aspergillus niger*, homologous constitutive expression, acidophilic, halophilic, homology modeling

# 529580 Adaptive Evolution of Microalgae *Schizochytrium* Sp. Under High Salinity Stress to Alleviate Oxidative Damage and Improve Lipid Biosynthesis

**Abstract:**

*Schizochytrium* sp. possesses a powerful lipid accumulation machinery that can be induced under stress conditions. However, stress-based strategies not only reduce the growth rate, but also cause oxidative damage to the cells, which can eventually lower the lipid yield. In the present study, adaptive laboratory evolution (ALE) combined high salinity was performed to enhance the antioxidant system and lipid accumulation. The final strain ALE150, which was obtained after 150 days, showed a maximal cell dry weight (CDW) of 134.5g/L and lipid yield of 80.14g/L, representing a 32.7 and 53.31% increase over the starting strain, respectively. Moreover, ALE150 exhibited an overall higher total antioxidant capacity (T-AOC) and lower ROS levels than the starting strain. Furthermore, the regulatory mechanisms responsible for the improved performance of ALE150 were analyzed by comparative transcriptomic analysis. Genes related to the transhydrogenase cycle were upregulated while the pentose phosphate pathway (PPP) pathway was down-regulated. Moreover, the metabolic fluxes towards the fatty acid synthase (FAS) and polyketide synthase (PKS) pathways were also changed. Interestingly, an overall increase of gene expression levels of superoxide dismutase (SOD) and catalase (CAT) was essential for the survival of *Schizochytrium* sp. under 30 g/L salt.

# 537992 Adapted Evolution and Biosensor-Based Screening for Robust Growth of *Pseudomonas Putida* on Corn Stover Hydrolysate and C*is,Cis*-Muconic Acid Production

**Abstract:**

Despite the advancements in synthetic biology tools, our capacity to assemble DNA parts for engineering novel pathways cannot be matched to the complexities of biological systems. Adaptive laboratory evolution for superior genotypes and phenotype physiology coupled with screening for superior variants using target metabolite biosensor can be a powerful technique to eliminate the bottlenecks in the engineering of biosynthetic pathways. In the search for alternate sustainable technologies for making products originally sourced from petroleum, lignocellulosic biomass are an abundant renewable source of sugars for microbial production of bio-based commodities. Unlike from starch, they do not compete with food supply as well. In the current study, *cis,cis*-muconic acid (ccMA) producing strain of *Pseudmonas putida* was continuously grown on sugar hydrolysate of corn stover and the resulting diverse population was screened using ccMA sensor for variants with enhanced growth and product titers.

# 530601 A High-Throughput Platform Technology for Engineering Enhanced-Solubility in Biotherapeutics

**Abstract:**

Biomolecular aggregation is a ubiquitous challenge encountered in the development of biotherapeutics. For manufacturers, protein aggregation can complicate biotherapeutic production, purification, formulation and storage, while for patients, biotherapeutic aggregates can induce unwanted immune responses. Enhanced solubility is therefore a desirable property for biotherapeutic engineering programs. Directed evolution using large combinatorial libraries can be a powerful strategy for solubilizing aggregation-prone biomolecules, but such an approach requires a robust and generalizable HT-screening method for protein aggregation. Here, we describe a platform technology for ultra-high throughput screening of protein solubility in *E. coli* expression hosts. Using high speed flow cytometry, the aggregation propensity of arbitrary protein libraries can be screened at rates of 10,000-30,000 clones per second, enabling the rapid isolation of highly soluble variants. By integrating deep sequencing into the experimental workflow, we have constructed a high-resolution map of the sequence-solubility-function space for a GFP model system, and we further demonstrate the utility of the platform with biotherapeutics of practical interest. This powerful aggregation screening method is quantitative, applicable to a vast array of protein targets, and enables rapid sorting of massive recombinant libraries. We anticipate that this technology will provide novel solutions to the aggregation problems suffered by many biotherapeutic agents.

# 532916 Well-Mixed Cancer-on-Chip System for the Simultaneous Evaluation of Toxicity and Efficacy of Anti-Cancer Drugs

**Abstract:**

Cancer research relies heavily on static 2D cell culture systems that do not recapitulate the complex physiology of solid tumors. Systems with greater physiological relevance are needed to expedite the testing of the efficacy and safety of novel anti-cancer drugs and to enable personalized medicine applications.

We developed and validated the use of a cancer-on-chip system that is based on the integration of a continuous 3D-printed mini-stirred tank and a portable incubator. This system allows the creation of homogeneous and well-controlled environments for the culture of cancerous 3D microtissues. We fabricated cancerous microtissues by bioprinting recombinant MCF7 cancerous cells (producers of green fluorescent protein) surrounded by human fibroblasts to mimic the coexistence of segregated areas of malignant and healthy cells (sharing surface) in solid tumors. These microtissues were continuously fed with culture medium in the mini-stirred tank and exposed to Docitaxel (as a model drug) for different residence times (ranging from 4 to 36 hours) and doses. Sustained or single injection additions of Docitaxel were tested. We characterized the response of cancerous/healthy microtissues by using fluorescence microscopy combined with immunofluorescence techniques, and RT-PCR to measure changes in cell viability, morphology, and expression of relevant biomarkers.

Our findings suggest that this cancer-on-chip system properly recapitulates the responses of cancerous and normal tissues to insults related to the administration of different concentrations of an anti-cancer drug.

# 522604 Modeling Chemical Transport in PDMS-Based Organ-on-Chip Microsystems

**Abstract:**

Organ-on-chip microsystems are being used to evaluate chemical responses to human cells cultured in appropriate 3D heterotypic environments. These microsystems are often fabricated from polydimethylsiloxane (PDMS), which has high affinity for small hydrophobic molecule creating undesirable change in chemical dose response curves and timing of exposure to cells. Here, we modeled the chemical adsorption onto PDMS surfaces in an organ-on-chip microsystem using computational fluid dynamics (CFD). The goal of this model is to predict the cultured cellsâ€™ actual time-dependent chemical exposure within a PDMS fabricated microsystem. It was possible to establish quantitative relationships for chemical adsorption onto PDMS surfaces through fitting spectroscopic data to a microscale model of chemical binding kinetics with PDMS and extracting time-dependent adsorption coefficients, saturation amount, and forward and reverse rate constants. Experimental kinetic rate constants were used to model adsorption due to continuous and bolus exposure of three chemicals and subsequent chemical transport in a hypothetical microsystem. These chemicals were hexazinone having no interaction with PDMS surfaces (the octanol-water partition coefficient, log P = 1.85), ethofumesate having reversible binding with PDMS surfaces (log P = 2.7), and propiconazole with strong affinity to PDMS surfaces (log P = 3.72). From this analysis, we determined that timing is critical for delivery of chemicals that reversibly bind to PDMS in order to avoid over- or under-dosing cells. For such chemicals, a bolus dose at the inlet may translate into an extended exposure for cells in the device due to delayed release of the chemical from PDMS surfaces. In addition, for chemicals with strong affinity to PDMS, the actual exposure may be an order of magnitude less than the nominal inlet concentration. The model can be used in a forward direction to properly evaluate the effect of chemicals by accurately quantifying cellular exposures or in a backward direction to achieve desired exposures by optimizing inlet dosing strategies.

# 531045 The Role of Industry in Helping Shape University Brewing Education Programs

**Abstract:**

In the last five years, an increasing number of universities have initiated academic brewing programs aimed at meeting the needs of the brewing and distilling industry—both at the "craft" and the industrial scale. Some programs very specifically try to fill gaps in knowledge of brewing science, while other programs focus on running a business. Still other programs are concerned with the industry as a whole.

These programs have "grown up" amid a decrease for public funding of higher education, as well as increasing public pressure to focus on immediate post-graduation employment. Whatever the merits of these two pressures, they are the reality, and brewing and distilling programs must address this.

The craft brewing and distilling industries play a key role in the Kentucky economy, and faculty at Western Kentucky University have expertise in various related areas. From 2010 up through 2014, representatives from the college of science and the college of arts and humanities interviewed brewers at the Great American Beer Festival in Denver and brewers and distillers around Kentucky to get a sense of industry needs in order to help shape a nascent brewing and distilling program.

The data from those interviews helped shape program curriculum into an interdisciplinary, workforce development program. The brewing and distilling program draws on expertise from the departments of chemistry, biology, history, management, entrepreneurship, tourism, agriculture, and others.

Those interviews also facilitated a corporate partnership that helped seed the program with more than $500,000 in cash and in-kind donations, including building renovation, a new student classroom-lab, and the installation on campus of a large production-level brewery staffed by our students. The new program is financially independent of the university, and the gift came with yearly financial program support and a licensing agreement on beer sales that provides an ongoing, independent revenue stream for the program. The collaboration led to benefits for other programs—the corporate partner now sends a yearly, fully-funded cohort to the MA in Agriculture, is exploring similar options with the Department of Biology, and provided a major gift of equipment to the university farm. Students with majors as diverse as Agriculture, Photojournalism, Chemistry, Tourism, and Economics now work in the industry, and the intercollegiate model has attracted national and international attention.

Dr. Andrew McMichael is the Associate Dean of the Potter College of Arts & Letters at Western Kentucky University, a Professor History, and the co-Coordinator of the Brewing and Distilling Arts & Sciences Program. This presentation will focus on the development of the Brewing and Distilling Arts & Sciences program at Western Kentucky University, and how corporate partnerships at the initial stages, as well as ongoing collaborations, have helped shape a workforce development program in brewing and distilling.

# 532250 On the Origin and Evolution of Brewing Science and Technology at Villanova

**Abstract:**

In 2017, Villanova introduced an elective course in Chemical Engineering titled "Brewing Science and Technology". Developed by Associate Professor Michael Smith, the course approaches the subject as the application of science and technology to the craft of brewing, much as a *craft* brewer would. Students brew beer using simple home-brewing equipment, and the history of brewing and brewing technology is covered with special emphasis on how various technologies (the thermometer, the microscope, refrigeration) altered the production, styles and flavor of beer we know today. This is followed by a deep dive into the biochemistry of malt, hops and yeast. As one student said, "I didn't know I was going to be dropped into the middle of an Organic Chemistry exam;" this aspect of the course qualifies it for Villanova's Minor in Bioengineering. The goal is not to train new brewers, but to illustrate how the mastery of this (or any other similar) enterprise requires a deep and nuanced understanding and application of chemical and biological sciences. This talk will cover the origins of the course, learning objectives and strategies, and future plans.

# 536253 Designing a Brewery Engineering Minor within Chemical Engineering to Meet MBAA Specifications

**Abstract:**

Much has been made about the natural overlap between chemical engineering and brewing: the fluid dynamics, the heat transfer, the fermentation kinetics, etc. Arguments have been made that beer production is a great case study for chemical engineering education and chemical engineers make great brewers. The last decade has seen an expansion of fermentation science/brewing programs, some as standalone programs and some within chemical engineering. Over the last two years, the development process for the Brewery Engineering program minor at NMSU has focused on how to add a minor to the chemical engineering curriculum to give its graduates an advantageous credential for employment in the brewing industry: a degree from a program recognized by the Master Brewers Association of the Americas (MBAA). In this presentation, we describe the design of new courses and selection of existing course to achieve this goal. We also describe our attempt to differentiate from general fermentation science programs by offering a more engineering perspective on design and operation of a brewery.

# 538757 From Concept to Class: Pitt's Engr 1933 – Engineering a Craft Brewery

**Abstract:**

The craft brewing industry is expanding rapidly: year-over year growth in the number of craft breweries has been over 15% annually since 2012 (http://brewersassociation.org). There are 6266 craft breweries operating at the start of 2018.. The industry provides an outlet for engineers looking to use their technical skills in brewing and distilling, as well as for entrepreneurs who may consider starting their own brewing business. Through this course, students can learn impactful problem-solving. Engineering skills, combined with solid scientific understanding, allow students to design solutions to production-oriented challenges.

The content of the course spans the brewing process, from reactants (malt, water, hops, and yeast) to packaged product. We introduce the brewing process and beer styles, followed by a step-by-step walkthrough of the equipment and operating conditions needed to design and yield a customer-pleasing product. The raw materials – base and specialty malts, water with specific chemistry, hops of various styles and flavors, and the breadth of yeasts – are discussed in the context of product design, where the students design beers we brew and sample at a semester-end event. Engineering calculations are introduced throughout the course to achieve the design specifications and move the product through the various units. Part of the presentation will focus on the course content, assessments and projects, and student perceptions.

Putting a class like this in place is decidedly more difficult than the usual engineering course, given the possible issues that can arise when mixing undergraduates and beer. The other focus of this presentation will be the legal, policy, and University challenges that we encountered in proposing ENGR 1933, and ways others could benefit from our experience. A key challenge was the intent to serve alcohol as part of the course, which was ultimately approved.

This class has been offered twice, once in 2017 and once in 2018. The first class was the most successful first-time offering of an elective in the history of the Swanson School of Engineering. The almost 150 students who have taken the class to date find it challenging, but very enjoyable. The presentation will close by discussing the design and execution of our 2018 study abroad trip: Farm to Tap in the Belgian Style. The overall aim of this presentation is to help those interested in launching a class like this on their campus to move quickly through the design and implementation hurdles that could otherwise slow their first offering.

# 541606 Brewing and Distilling: Alive and Well in Northwest Arkansas and the University

**Abstract:**

The inaugural brewing competition at AIChE came at an interesting time for the Northwest Arkansas community, for not only was there local interest in the chemical engineering program to demonstrate skills associated with beer manufacturing, at the same time general interest prompted the University of Arkansas to develop and deliver a certificate program to serve this growing market. This presentation will begin with a synopsis of the experience our student team had in an environment with little infrastructure, and end with the plans to offer â€œBrewing and Distillingâ€ based on the desire to provide this certificate to people interested in the craft beer industry that has exploded in our area.

# 533485 Optimization of Aroma Profiles through Selective Removal of Off-Flavors: An Exemplary Study in Alcohol-Free Beers

**Abstract:**

A delicate balance of aroma molecules characterizes the distinctive flavor profile of foodstuffs and beverages. If off-flavors are present, this profile is distorted. In alcohol-free beers, such a lack of balance is often induced by the enhanced perception of wort flavors. In general it is recognized that wort flavor is caused by so-called Strecker aldehydes that are typically produced by a heat-induced reaction between a reducing sugar and an amino acid. Nonetheless, formation may occur via multiple routes, therefore their occurrence is difficult to prevent. The aim of this work is hence to develop a technology that can engineer the desired flavor profile and fine-tune the aroma concentration to the respective product. Considering the sensitive nature of foodstuffs and beverages, a high selectivity at mild operation conditions is required, desirably with low capital investment and operational costs.

In the first part of this study, a technical solution to this challenging task was developed by identifying potential bottlenecks and proposing process concepts circumventing these. Consequently, the feasibility of the technology was evaluated to demonstrate the proof of concept. In the final step, the impact of the beverage constituents on the process efficiency was tested to gain insight into the involved mechanisms and to determine process robustness. Based on these findings, it is possible to design a new processing operation that facilitates a controlled off-flavor capture from alcohol-free beers.

# 539283 Functionalized Mesh Materials for Listeria Control in Dairy Applications

**Abstract:**

Pathogenic bacteria are ubiquitous in dairy systems. They are mostly eliminated by pasteurization, though improper treatment and recontamination can occur during downstream processing. Spiral elements are used widely in downstream processing of milk, and offer an excellent opportunity for mitigation of pathogens. Specifically, the feed spacer in spiral elements has been functionalized to provide anti-microbial activity. The surface may be further modified to target specific pathogenic bacteria. The purpose of this paper is to discuss development of these materials targeted for control of *Listeria Monocytogenes* (LM) with applications in dairy processing, including milk and brine systems.

The feed spacer mesh has been modified to include anti-microbial groups. Polypropylene (PP) feed spacer has been modified in several steps to obtain several types of quaternary amine (QA) groups. The QA groups are on polymer grafts attached to the PP spacer. Polymerization of the base polymer was achieved by photo-polymerization, and functional groups were added post-polymerization. We have characterized the polymerization kinetics as a function of time, monomer concentration, and initiator concentration. We have also demonstrated anti-microbial activity against Gram-positive bacteria. We will discuss our results so far, as well as methods to incorporate pathogen-selectivity.

# 530434 Electrospun Carbon Nanotube/Sulfonated Poly (ether ether ketone) Proton Conductive Membranes for Vanadium Redox Flow Battery

**Abstract:**

Vanadium redox flow battery (VFB) is a promising large-scale energy storage techniques for most renewable energy resources. As one of the key components in VFB, proton conductive membrane faces the challenge of improving selectivity of proton to vanadium ions (H/V selectivity), because both vanadium ions and proton transport through the similar pathway of hydrophilic domains in polymer matrix. Inorganic additives has intrinsically good vanadium ions barrier, while are likely to agglomerate or peeling away from the polymer materials due to poor compatibility. In this work, carbon nanotube (CNT) hybrid sulfonated poly (ether ether ketone) (SPEEK) is nano-fiberized through electrospinning. High electric field inducesthree-dimensional interconnected SPEEK nanofiber networks and thealignment of fibrous CNTs along SPEEK nanofiber axis, which provide long-range proton conductive pathways,good dispersion of CNT in SPEEK nanofibers and maximum vanadium ions and swelling blocking effect. Area resistance of Spun 0.5 is reduced to 0.23 ohm/cm$^2$, which is 20.7% lower than that of Cast0.5.Vanadium ions permeation decreases by about 67.6% at CNT content of 0.5 wt% comparing to 0 wt.%. Good H/V selectivity and tensile strength in hydrated state make the electrospun MWCNT/SPEEK membrane exhibit good VFB performance.At current density of 100 mA cm$^{-2}$, energy efficiency of Spun0.5 is about 86.2%, much better than that of Cast0.5 (77.7%) and Nafion211 (74.9%). Even after 100 cyclic operation, Spun0.5 shows higher capacity stability (0.47% per cycle decay), which is much better than that of Nafion 211 (0.84% per cycle decay) with similar membrane thickness.

Journal of Membrane Science 555 (2018) 268–279

Contents lists available at ScienceDirect

Journal of Membrane Science

journal homepage: www.elsevier.com/locate/memsci

# Directing filtration to narrow molecular weight distribution of oligodextran in an enzymatic membrane reactor

Ziran Su[a,b], Jianquan Luo[a,b,*], Manuel Pinelo[c], Yinhua Wan[a,b,*]

[a] State Key Laboratory of Biochemical Engineering, Institute of Process Engineering, Chinese Academy of Sciences, Beijing 100190, PR China
[b] University of Chinese Academy of Sciences, Beijing 100049, PR China
[c] Department of Chemical and Biochemical Engineering, Technical University of Denmark, 2800, Kgs, Lyngby, Denmark

ARTICLE INFO

*Keywords:*
Ultrafiltration
Oligosaccharide
Dextranase
Membrane fouling
in-situ product separation

ABSTRACT

Oligodextrans with molecular weight (Mw) within the range of 5.0–8.0 kDa have great commercial potential as precursors of iron-dextran for anemia treatment. Traditional oligodextran production consists of sucrose fermentation, acid hydrolysis and ethanol precipitation, which results in an uneven Mw product, hypersaline wastewater discharge and potential safety hazards. In this work, a novel enzymatic membrane reactor (EMR) system to produce oligodextran is proposed, whereby in-situ product recovery can be manipulated to control the Mw distribution of the resulting products. Results showed that the membrane material played an important role in the permeate flux and transmission of oligodextran. Among the tested membranes, a 20 kDa polyethersulfone (PES) membrane was found to be optimal for building up the EMR, as it successfully controlled the oligodextran Mw within the desired range with a relatively narrow distribution and high productivity. Moreover, high transmembrane pressures (3 bars) and low stirring rates (160 rpm) promoted yields beyond 50% in 120 min. Higher permeate fluxes prevented further product hydrolysis and enhanced the yield. However, the resulting concentration polarization (CP) should be minimized to reduce accumulation of large oligodextran molecules on the membrane surface, which might diffuse through the membrane and thus broaden the Mw distribution of the products in the permeate. Both dextranase and dextran caused membrane irreversible fouling. The fouling caused by the enzymes not only favored the enzyme immobilization itself, but also contributed to narrow the membrane pore size distribution. As a result, a higher uniformity of oligodextran products compared with the pristine EMR was obtained, especially at the beginning of operation with EMR (which was improved by 22%). It was concluded that selecting the suitable membrane type and permeate flux, maximizing the shear rate, and narrowing the membrane pore size distribution were effective strategies to obtain high-quality oligodextran products by EMRs.

## 1. Introduction

Linked by alpha-1,6 glucosidic bonds, dextran is one of the most known polysaccharides made of repeating glucose units and is widely used in food and medicine [1]. Dextran molecules exhibit particular properties depending on the molecular weight (Mw). Especially oligodextran with certain Mw (5.0–8.0 kDa) has superior properties for maintaining blood pressure, extending blood vessels and for being able to form iron complexes for clinical usage [1–3]. Traditional oligodextran production consists of sucrose fermentation, acid hydrolysis and ethanol precipitation. *Leuconostoc mesenteroides* converts sucrose to macrodextran molecules with Mw over several million Dalton [4–6]. Acids are thereafter used to break the alpha-1, 6 glycosidic bonds randomly, generating smaller pieces of dextran. Several alcohol sedimentation steps are subsequently required for purifying oligodextrans at several Mw ranges after acid hydrolysis [7]. Such method can result in the occurrence of chlorinated species amongst the oligodextran products, which may cause serious damage to human beings. Additionally, considerable amounts of ethanol are consumed during oligodextran purification, which has to be recovered and recycled. More importantly, the obtained oligodextrans have a wide distribution of Mw, which may result in immune problems in clinical usage [8].

Production of oligodextran by enzymatic hydrolysis can be considered safer, as it occurs at milder conditions and involves less wastewater generation. Dextranase from *Penicillium sp.* can effectively break the alpha-1,6 glycosidic bonds of the dextran molecules under mild aqueous environment [9]. No extra chemicals need to be added to the reaction system, which also contributes to make the process

* Corresponding authors at: State Key Laboratory of Biochemical Engineering, Institute of Process Engineering, Chinese Academy of Sciences, Beijing 100190, PR China.
*E-mail addresses:* jqluo@ipe.ac.cn (J. Luo), yhwan@ipe.ac.cn (Y. Wan).

https://doi.org/10.1016/j.memsci.2018.03.062
Received 26 January 2018; Received in revised form 20 March 2018; Accepted 22 March 2018
Available online 23 March 2018
0376-7388/ 

greener. While studying a dextransucrase–sucrose system (synthesis of dextran), Gan et al. found that an early addition of dextranase resulted in the production of low Mw of oligodextran products [8]. Likewise, Iqbal and co-workers reported that dextranase hydrolysis generated less dispersed oligodextran compared to acid hydrolysis, which made oligodextran production of clinical grade possible [10]. Additionally, Goulas et al. demonstrated that low enzyme concentrations could generate narrower oligodextran distributions during enzymatic hydrolysis [11].

Besides optimization of the enzymatic reaction, membrane technology, especially ultrafiltration (UF), has been previously used to narrow the Mw distribution of oligodextran due to its low energy requirement and sustainability [12]. Early in 1999, an UF membrane was used to separate certain clinical oligodextran products after enzymatic hydrolysis [13]. Later research further developed a continuous membrane reactor in which Mw distribution could be controlled by adjusting enzyme concentration and retention time [14]. Then Torras and co-workers optimized the dextranase loading to be immobilized on a polymeric membrane for controlling the products Mw [15]. Pu et al. also found that a UF-diafiltration process could greatly narrow the Mw distribution of clinical dextran [16]. In most of the previous studies, however, enzymatic hydrolysis and UF were performed separately. Actually, enzyme reuse and in-situ product recovery could be carried out simultaneously by coupling enzyme and membrane technology in the same system (Fig. 1), an enzymatic membrane reactor (EMR). Among other advantages, such solution could probably reduce product inhibition of the enzyme and avoid further hydrolysis of target products. Previous studies on the topic involving the use of EMR mainly focused on the optimization of enzymatic hydrolysis rather than paying attention to the membrane separation process, which is indeed the product quality-limiting step. During membrane filtration, concentration polarization (CP) is considered an obstacle, resulting in the decrease of permeate flux and selectivity [12,17]. However, the increase in concentrations of substrate and enzyme on membrane surface could also favor solute transmission through the membrane, hence increasing enzymatic reaction efficiency. Membrane fouling, however, could not only decrease permeate flux and thus increase the retention time of substrates, but also might reduce the transmission of target products through the membrane [18].

In present work, aiming at producing oligodextran with Mw between 5.0 and 8.0 kDa at a narrow Mw distribution, an EMR was designed to enable enzyme reuse and in-situ product recovery. Large dextran molecules with initial Mw of 40 kDa were first hydrolyzed by dextranase. The effect of the membrane material, molecular weight cut-off (MWCO), transmembrane pressure (TMP) and shear rate (i.e. agitation speed) were evaluated in terms of oligodextran Mw and its distribution in permeate, permeate flux, irreversible fouling and yield. Since membrane fouling is inevitable during membrane separation, based on the understanding of fouling mechanisms of the established EMR, dextranase was partly immobilized on the UF membrane in advance by a simple filtration ("fouling-induced" enzyme immobilization [19,20]). We hypothesized that the "enzyme fouling" layer could not only increase enzyme stability and accelerate online product separation, but also narrow the distribution of membrane pore size (larger pores are more prone to be blocked since the higher local flux around the larger pores would induce more serious fouling formation) [21,22] and improve the separation selectivity.

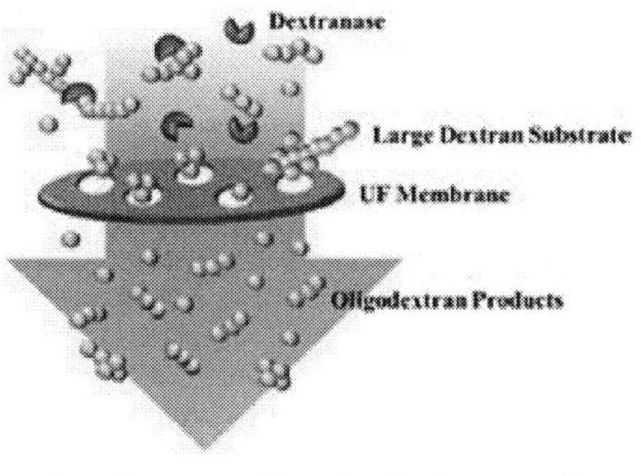


**Fig. 1.** Illustration of processes taking place in the enzymatic membrane reactor. Dextranase breaks the alpha – 1,6 glucosidic bonds on dextrans randomly and generates smaller dextran molecules with different molecular weights. A membrane with a particular pore size distribution is used under certain pressure to separate the target molecules while retaining large dextran fractions and enzymes. Target oligodextrans (5.0–8.0 kDa) are removed from the reactor to avoid further hydrolysis.

## 2. Methods and materials

### 2.1. Materials

Dextran substrate (DXT40K) with Mw 40 kDa was purchased from Sinopharm Co Standard. Dextran Benchmark with Mw 0.36, 5, 12, 25, 50, 80 and 150 kDa, dextranase (EC 3.2.1.11, dry powder from *Penilisilium. Sp.*) and Bradford reagent used for the protein assay were purchased from Sigma-Aldrich Co. The ultrapure water used for dissolving substrate and enzymes was generated by a Millipore Milli-Q Advantage A10 water purification system (pH 6.0 ± 0.5). Other chemicals were of analytic grade. Information of UF membranes with different materials and MWCO used in the experiments is listed in Table 1.

### 2.2. Experimental procedure

Enzymatic hydrolysis reactions were firstly carried out in beaker to obtain preliminary information about the Mw variations of dextran under different dextran/dextranase ratios (25 °C, 160 rpm agitation, pH 6.0 ultrapure water as solvent, chemical dosage was listed in Table S1). 0.5 ml samples were collected every 30 min and placed in a boiling water bath for 1 min to deactivate the enzymes.

After a suitable dextran/dextranase ratio was selected, an EMR was constructed and operated at room temperature (25 °C). The fresh membranes were immersed into 5% NaCl (regenerated cellulose membranes) or 0.01 M NaOH (other membranes) solution for 30 min to remove the preservatives, according to the instructions of the manufacturers. The clean membranes were then placed at the bottom of a dead-end stirred cell (Amicon 8050, Millipore, U.S.A) with an effective membrane surface area of 13.4 $cm^2$ or a home-made cell with an effective membrane surface of 4.52 $cm^2$, standing above the magnetic stirrer. A constant pressure was generated by filling nitrogen gas into the cell while a constant flux was provided by LC-20AT pump (SHIMADZU). The schematic diagram of these two systems was shown in Fig. S1. The membrane was compacted at 4.5 bar for 30 min with pure water. Dextran substrate and dextranase were mixed in the cell which was continuously fed with pure water under different pressure and agitation speed. A cylinder was used for measuring and collecting the permeate samples every 30 min. All the collected samples were heated in boiling water for 1 min and then stored at 4 °C for analysis.

For the experiments dealing with the study of fouling mechanisms,

**Table 1**
Properties of membranes used in the experiments.

| Membrane type | Skin layer material | MWCO (Da) | Manufacture |
|---|---|---|---|
| RC5 | Regenerated cellulose | 5000 | Millipore |
| RC10 | Regenerated cellulose | 10,000 | Alfa Laval |
| PES10 | Polyether sulfone | 10,000 | AMFOR |
| PES20 | Polyether sulfone | 20,000 | AMFOR |
| PES30 | Polyether sulfone | 30,000 | AMFOR |
| PAN10 | Polyacrylonitrile | 10,000 | AMFOR |

50 ml 0.05 g/L dextranase and 50 g/L 20 kDa dextran (Sinopharm Co.) were respectively added into the filtration cell equipped with PES20 membrane. Moreover, 50 g/L DXT40K was hydrolyzed by 0.05 g/L dextranase for 30 min, 60 min, 90 min and 180 respectively. After reaction, the samples were heated in boiling water for enzyme deactivation and then added into the filtration cell where the PES20 membrane was placed. For these filtrations (each feed was 50 ml), pure water was continuously fed into the cell under 4.5 bar and 2000 rpm for 2 h. Water permeability before and after filtration was measured in order to assess the irreversible fouling.

For the "fouling-induced" enzyme immobilization process, a fresh membrane was placed at the bottom of the filtration cell, and 2.5, 5.0 and 10 mg dextranase in 50 ml pure water were added respectively. The filtration was conducted in concentration mode under 4.5 bar and 100 rpm. After 48 ml of solution permeated through the membrane, the filtration was stopped. 0.5 ml of retentate and permeate were respectively collected for protein concentration determination. The rest of retentate and permeate were mixed with 50 ml 50 g/L dextran substrates for the hydrolysis and separation in this special EMR under 4.5 bar and 2000 rpm.

### 2.3. Analysis methods

Gel permeation chromatography (GPC) was used to test the average Mw of products and their distribution. Samples were eluted under 1 ml/min in ultrapure water, 40 °C. A refractive index detector coupled with the G4000PWXL column from Shimadzu was used for testing the samples. Peaking information including Mw and dispersity (D = Mw/Mn, Mw is weight-average molar mass, Mn is number-average molecular weight) of each sample were determined with help of the Shimadzu LC solution software. A smaller D value represents a higher uniformity of oligodextran products. The benchmark of the dextran molecules in the range between 0.36 and 150 kDa is given in Fig. S2. The average Mw was the arithmetic mean Mw of the four collected samples during the 120 min of hydrolysis.

The phenol-sulfuric acid method was used for testing the total saccharides concentration in the permeate for the different enzymatic systems. Samples were diluted by 500 times with pure water before testing. Each 2 ml diluted sample was added to a 10 ml glass tube. 1 ml of 6% (v/v) phenol and 5 ml of sulfuric acid were mixed with the samples. Then the mixtures were shaken for 30 s and let stand for other 30 min. The sample absorbance was tested in proper cuvettes under 490 nm in a UV-2100 Spectrophotometer. Glucose solutions with concentrations between 0.01 and 0.07 g/L were used for calibration (Fig. S3a). The obtained saccharides concentrations were the average saccharides concentrations corresponding to each time interval (0–30 min, 30–60 min, 60–90 min and 90–120 min).

The enzyme activity and oligodextran production rate were tested by using 3, 5-dinitrosalicylic (DNS) acid reagent. 5 mg free dextranase in 30 ml $H_2O$ was stirred under 160, 1000 and 2000 rpm (25 °C) respectively. 1 ml of enzyme solution was taken every 30 min, and then mixed with 10 ml 4 g/L DXT40K for 25 min followed by deactivation in boiling water. Reducing sugar concentration in the solution was measured by using DNS reagent at 540 nm. The enzyme activity was defined as the amount of isomaltose (measured as maltose with unit of umol) per unit of enzyme mass (mg) after 1 min at 25 °C. In the oligodextran production rate testing, 4 g/L DXT40K solution (50 ml) was mixed with 0.5 ml 0.2 g/L dextranase. The reactions were performed for 2 h under 160, 1000 and 2000 rpm, respectively (25 °C). The oligodextran production rate was defined as the cumulative maltose amount (umol) per unit of enzyme mass (mg) at corresponding reaction time interval. The reducing sugar was tested by the same method mentioned above. Maltose solutions with concentrations between 0.2 and 0.8 g/L were used for calibration (Fig. S3b). The protein concentration in enzyme solution was measured by Bradford assay with a spectrophotometer at 595 nm (Fig. S3c).

### 2.4. Characterization of membrane

The morphology of pristine and "fouled" PES20 membranes was characterized by field emission scanning electron microscopy (FESEM, Hitachi S480, Japan) with an acceleration voltage of 5.00 kV after the samples being sprayed with a thin gold layer (around 20 nm). Membrane surface porosity and pore size distribution were calculated from SEM images by Image J software.

### 2.5. Calculations

#### 2.5.1. Membrane water permeability and permeate flux

Permeate flux ($J_p$) was calculated as:

$$J_p = \frac{V_p}{t \cdot A_m} \tag{1}$$

Membrane water permeability ($L_p$) was calculated by the following equation:

$$L_p = \frac{\mathbf{J_p}}{\mathbf{TMP}} \tag{2}$$

where $V_p$ is the permeate volume corresponding to certain time (t) during the testing, $A_m$ is the effective filtration area, TMP is the transmembrane pressure.

The average permeate flux in an EMR was measured at four reaction time intervals (0–30 min, 30–60 min, 60–90 min and 90–120 min) respectively and calculated by Eq. (1).

Irreversible fouling was defined as the percentage of water permeability loss:

$$F_{irr} = \frac{\mathbf{L_{pi}} - \mathbf{L_{pa}}}{\mathbf{L_{pi}}} \times 100\% \tag{3}$$

where $L_{pi}$ and $L_{pa}$ are the $L_p$ before and after filtration, respectively.

#### 2.5.2. EMR yield

The volume of permeate samples was measured every 30 min. The amount of saccharides in each permeate sample was calculated from the saccharides concentration ($C_p$) and volume in the permeates:

$$M_p = C_p \times V_p \tag{4}$$

In the water feeding system, the initial mass of dextran was settled, and the yield was calculated as:

$$Y = \frac{M_p}{Mt} \tag{5}$$

where $M_p$ is the sum of saccharides mass in permeate, and $M_t$ is the total mass of saccharides added in the reaction system.

#### 2.5.3. Enzyme loading efficiency

The immobilized amount of enzyme was calculated by

$$M_{imm} = C_{ini.} \times V_{ini.} - C_{per.} \times V_{per.} - C_{ret.} \times V_{ret.} - C_{rin.} \times V_{rin.} \tag{6}$$

where $M_{imm}$ is the immobilized mass of enzyme, $C_{ini.}$, $C_{per.}$, $C_{ret.}$ and $C_{rin.}$ represent enzyme concentrations in the feed, permeate, retentate and rinsing solution respectively. V is the volume of the solutions corresponding to the concentration.

The loading efficiency is defined as:

$$\text{Loading} = \frac{M_{imm}}{C_{ini.} \times V_{ini.}} \times 100\% \tag{7}$$

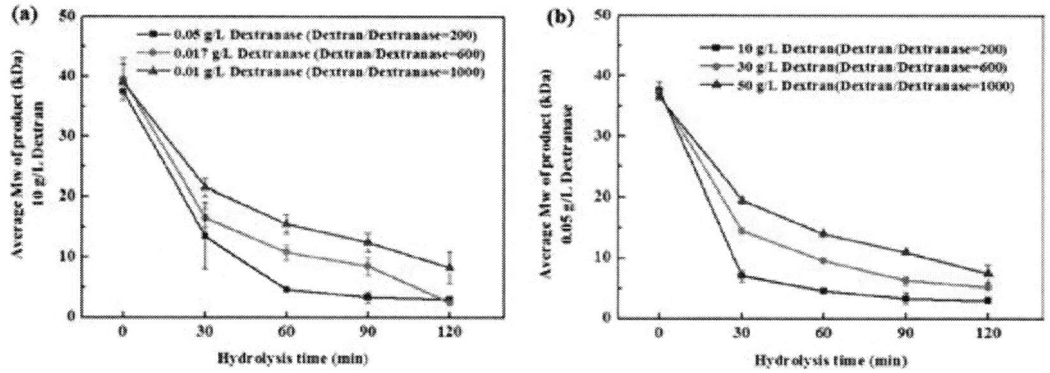


**Fig. 2.** Effect of concentrations of dextran and dextranase on the average Mw variations in a batch hydrolysis system. (a) 10 g/L dextran substrate with different dextranase concentrations, (b) 0.05 g/L dextranase with different dextran concentrations.

## 3. Results and discussion

### 3.1. *Optimizing operational variables in the EMR*

#### 3.1.1. *Effect of substrate and enzyme ratio*

Within certain reaction time, substrate and enzyme concentrations are the main variables affecting the average Mw of products [23]. This was confirmed by the results in Fig. 2, where it can be seen that larger dextran/dextranase ratios resulted in slower decreases of Mw in the batch reaction system. For a low dextran/dextranase ratio of 200, the average Mw declined fast within the first 30 min from 37.5 ± 1.5 to 7.1 ± 1.0 kDa and slowed down from 30 min to 120 min. The average Mw of the final oligodextran product in this case was 3.0 ± 0.2 kDa after 120 min hydrolysis, which was lower than the desired one (5.0–8.0 kDa).

The main goal of this work was to establish an EMR that can carry out simultaneous product separation during hydrolysis. For that purpose, the production rate of target oligodextrans within Mw 5.0–8.0 kDa should match the rate of product removal of the membrane (i.e. reaction and separation should be synchronous). If the product filtration is not quick enough, the products would be further hydrolyzed to smaller pieces, leading to product quality decline. In order to better match the kinetics of the enzymatic reaction and the filtration, a high dextran concentration (50 g/L) and a high ratio of dextran/dextranase (1000) were chosen for the following experiment.

Based on the above results, 50 g/L dextran (DXT40K) together with 0.05 g/L dextranase were mixed in the stirred cell equipped with a PES20 membrane and continuously fed with pure water (pH 6.0 ± 0.5) under 3 bar. The average Mw of the products collected in the permeate side was 7.6 ± 0.2 kDa, which fell within the desired range (5.0–8.0 kDa) but was much lower than the nominal MWCO of the PES20 membrane. This was probably due to the pore narrowing and blocking caused by fouling formation and membrane compaction [24].

Because of the randomness of enzyme hydrolysis, it is possible that the rejected target oligodextrans might be further degraded by the enzymes leading to an undesired wide Mw distribution in the permeate at the beginning (Fig. S4). An attempt to incorporate a continuous dextran substrate feeding was then made to prevent further hydrolysis of target oligodextran products, also taking into account that dextranase is prone to hydrolyze larger dextran molecules [7]. However, as shown in Fig. S4, continuous substrate feeding under a constant TMP resulted in more accumulation of large dextran molecules, which in turn caused a severe CP layer on the membrane surface. As a consequence, more diffusion of large dextran molecules through the membrane occurred, which led to an even wider Mw distribution of products. From these results it was deduced that the strategy of controlling products average Mw and narrowing down their distribution should be rather focused on increasing the membrane transmission and selectivity, in order to ensure the removal of the desired products as soon as possible and to obtain maximum rejection of large dextrans. Therefore, membrane pore size and materials which are mainly responsible for the filtration behavior were further investigated to optimize this online separation process.

#### 3.1.2. *Effect of MWCO and membrane materials*

UF membranes with MWCO of 5–20 kDa and different skin layer material were screened to control the average Mw and distribution of products. In general, all the EMRs decreased the average Mw and distribution of products as compared to batch reaction system (Fig. 3a, b), however the results could not only be explained by the MWCO differences among the membranes. The RC5 membrane with the smallest pore size had the lowest water permeability and permeate flux (Fig. 3c, d). As a result, it produced the most uniform and smallest oligodextran molecules (1.7 ± 0.1 kDa). However, with the same nominal MWCO, the average Mw of products and their Mw distribution in the PES10-EMR were much lower than those in the RC10-EMR and PAN10-EMR, which was even close to those in the PES20-EMR. Such results suggested that the membrane material rather than the MWCO was the most determining factor during product separation for the membranes with larger pore size. It was speculated that the oligodextran molecules with abundant hydroxyl groups had stronger interaction with the hydrophilic RC and PAN membranes, inducing more transmission of large dextrans through the membrane [25]. On the other hand, the permeate flux in the PAN10-EMR was much lower than that in RC10-EMR and PES10-EMR (Fig. 3c). It is well known that the permeate flux is governed by both membrane and fouling resistances [18]. The PAN10 membrane had the highest permeability (lowest membrane resistance), and the irreversible fouling detected (permeability loss after filtration) was only slightly higher than the others (Fig. 3d). This implied that the CP or reversible fouling in the PAN10-EMR was most serious, which was caused both by the accumulation of dextrans on the membrane surface and by the strongest interaction between dextran and PAN membrane material which is rich in carboxyl groups (nitrile groups would be hydrolyzed to carboxyl groups). The larger average Mw of the resulting products in permeate and the wider Mw distribution for the PAN10-EMR than those for other EMRs (Fig. 3a, b) supported the above hypotheses.

The average Mw of the products from PES20-EMR was right within the desired range (7.6 ± 0.2 kDa), while the Mw distribution decreased from 3.59 at 30 min to 2.50 at 120 min. Moreover, the permeate flux for the PES20-EMR during filtration was the highest. Such high flux indicated that the PES membrane (less hydrophilic than the others) could not only reject more large oligodextran molecules resulting in a relative low average Mw and narrow distribution of the products, but

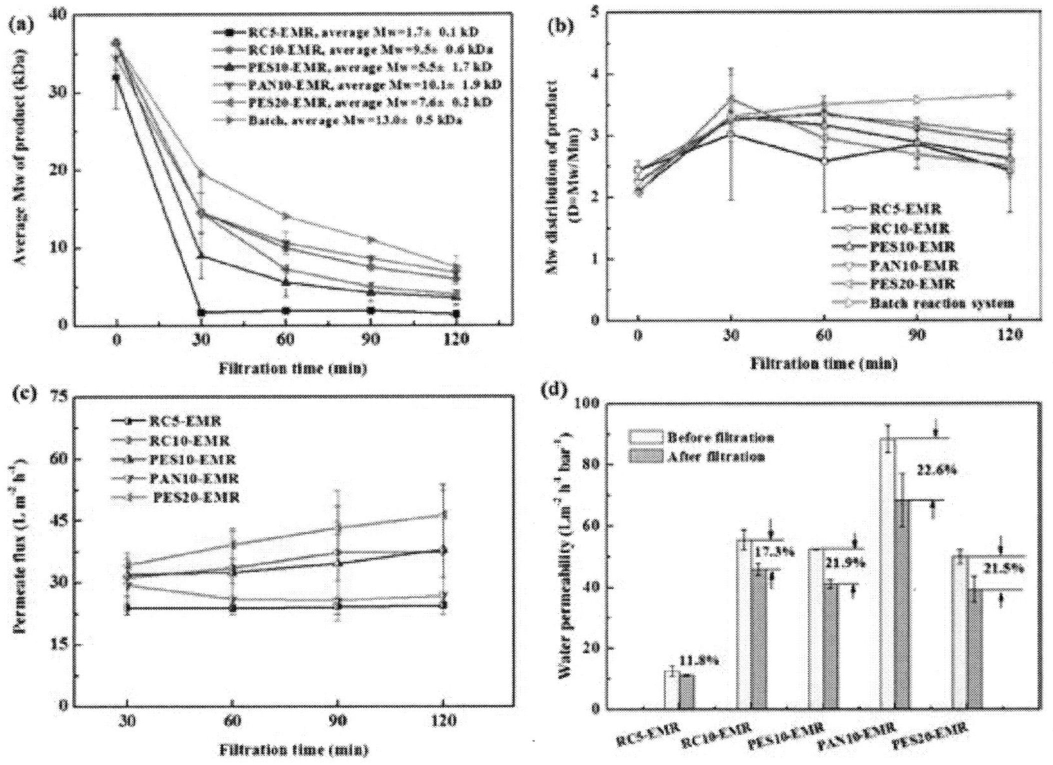


**Fig. 3.** Effect of membrane materials and MWCO on the performance of EMRs. (a) Average Mw, (b) Mw distribution of oligodextran molecules in permeate, (c) permeate flux with time and (d) irreversible fouling after filtration for different EMRs.

also had a thinner CP layer on the membrane surface due to the weaker interaction with dextrans (leading to a higher permeate flux). Therefore, the PES20-EMR outperformed the others in terms of product quality and permeate flux, which was selected for the further studies.

### 3.2. Process optimization

Resistance-in-series model is widely used in describing UF process [18,26] which can be expressed as follows:

$$J = \frac{TMP - \Delta\Pi_b}{\mu(R_m + R_{rf} + R_{if})} \tag{8}$$

Where $\Delta\Pi_b$ is the osmotic pressure difference between the bulk and the permeate (Pa), $\mu$ is the solvent viscosity (Pa s), $R_m$ is the membrane resistance ($m^{-1}$), $R_{rf}$ and $R_{if}$ are the resistances caused by reversible fouling and irreversible fouling respectively ($m^{-1}$).

The permeate flux is proportional to the driving force and negatively correlated to the viscosity as well as filtration resistances. Theoretically, higher TMP can increase the permeate flux until the limiting flux is obtained. Higher permeate flux would first produce more solvent convection, resulting in higher solute rejection (dilution effect), but further increasing permeate flux may cause severe CP and membrane fouling, leading to unexpected results. On the other hand, the higher shear rate induced by agitation on the membrane surface is expected to attenuate the CP and membrane fouling, resulting in higher permeate fluxes and solute rejections under constant TMP mode [27–29]. These theories have been intensively studied in the single membrane filtration. Since the filtration behavior also affects the enzymatic reaction efficiency, it is far more complicated to study such effects in an EMR.

#### 3.2.1. Effect of TMP

Fig. 4 illustrates effect of TMP on oligodextran production in a PES20-EMR at 160 rpm. A low pressure (1 bar) resulted in the smallest average Mw (4.9 ± 0.7 kDa) and also the most uniform products (Fig. 4a, b). Long retention time due to low permeate flux (Fig. 4c) enabled sufficient hydrolysis of substrate, thus lowering the product Mw. On the other hand, a low permeate flux produced the least irreversible fouling (Fig. 4d), and thus the effect of fouling layer on oligodextran transmission through the membrane was negligible, resulting in an increasing oligodextran concentration in permeate with time (Fig. 4e). However, the low permeate flux also limited the oligodextran yield, and the lowest yield was obtained at 1 bar after operation for 120 min (Fig. 4f). With increase of TMP, the permeate flux in the EMR was greatly improved, leading to severer CP [30] and shorter retention time. Accordingly, less hydrolysis time and stronger concentration-gradient diffusion across the membrane at higher TMP produced a product with higher average Mw (7.6 ± 0.2 kDa at 3.0 bar, 9.5 ± 0.3 kDa at 4.5 bar) and wider Mw distribution (Fig. 4a, b). The formation of a more prominent CP layer at higher TMP might induce two opposite effects on oligodextran transmission through the membrane: first, more accumulation of oligodextran in the CP layer could accelerate its diffusion across the membrane; second, a more serious CP could result in a higher irreversible fouling (i.e. pore narrowing and blockage), thus increasing oligodextran rejection. The former was more obvious at the beginning of the operation and at low TMP ranges, and could explain why the saccharides concentration in the permeate was increasing with TMP at 30 and 60 min. The effect of irreversible fouling on oligodextran transmission was dominant at 90 and 120 min, which led to lower saccharides concentration in the permeate at 4.5 bar (Fig. 4e). Finally, since the yield was calculated by multiplying permeate volume and saccharides concentration then dividing total

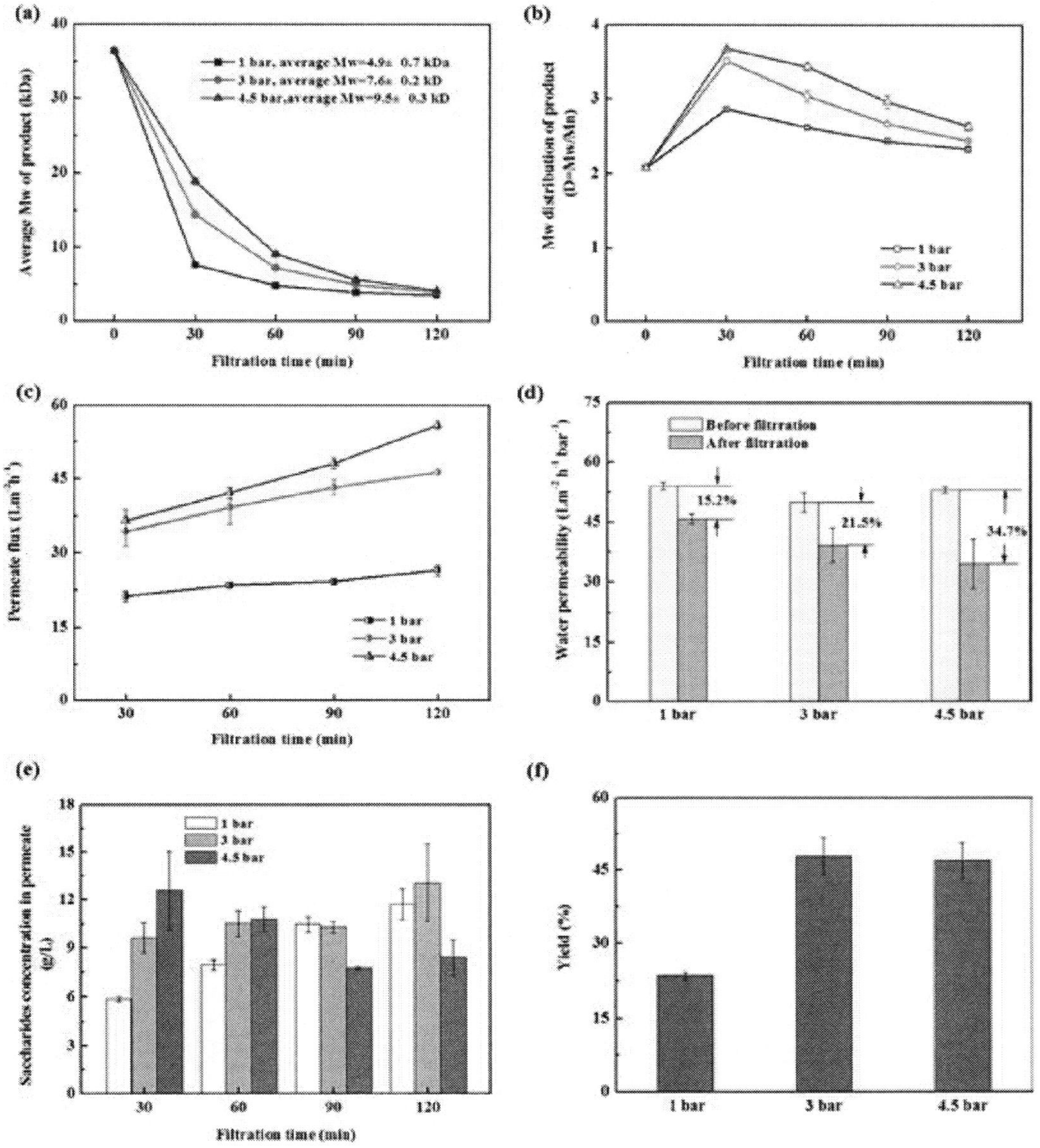


**Fig. 4.** Effect of TMP on oligodextran production in a PES20-EMR at 160 rpm. (a) Average Mw, (b) Mw distribution of oligodextran molecules in permeate, (c) permeate flux with time, (d) irreversible fouling, (e) saccharides concentration in permeate and (f) yield of oligodextran in permeate.

added saccharides amount (Eq. (5)), it became almost the same at 3.0 and 4.5 bar (Fig. 4f), which was much higher than that at 1.0 bar because of its lowest permeate flux (Fig. 4c).

It is worth mentioning that under constant TMP mode, the permeate flux was changing which might affect the validity of the above discussion. Thus, the effect of permeate flux on the oligodextran production in a PES20-EMR was investigated under constant flux mode. Based on the results in Fig. 4**c**, the permeate flux of 16, 27 and 40 $Lm^{-2}h^{-1}$ was applied. As shown in Fig. S5, the higher permeate flux resulted in higher product Mw and wider Mw distribution which was similar to the results in Fig. 4. However, the initial pressure was quite high and then decreasing under constant flux mode. The permeability loss was 24.1%, 31.3% and 56.5% at 16, 27 and 40 $Lm^{-2}h^{-1}$ respectively, indicating that severer irreversible fouling was formed at higher permeate flux. Although the CP-induced diffusion enhancement still existed, this denser fouling layer and higher permeate flux (dilution effect [18]) would increase the oligodextran rejection, thus decreasing the saccharides concentrations in permeate at higher permeate flux. Thus, the product yield was not improved when the permeate flux increased from 16 to 27 $Lm^{-2}h^{-1}$. Actually, the permeate flux should be a little lower at the beginning of the reaction because the target product was few while the CP should be controlled, and then the permeate flux ought to be increased when more oligodextran molecules were generated. Therefore, constant TMP mode resulting in an increase of flux with time was considered as "self-regulation" process, which might be more suitable for this application.

According to the above discussion, a higher TMP was desirable in order to avoid further product hydrolysis and promote high yield, but CP should be minimized to reduce accumulation of large oligodextran molecules on the membrane surface, which might diffuse through the

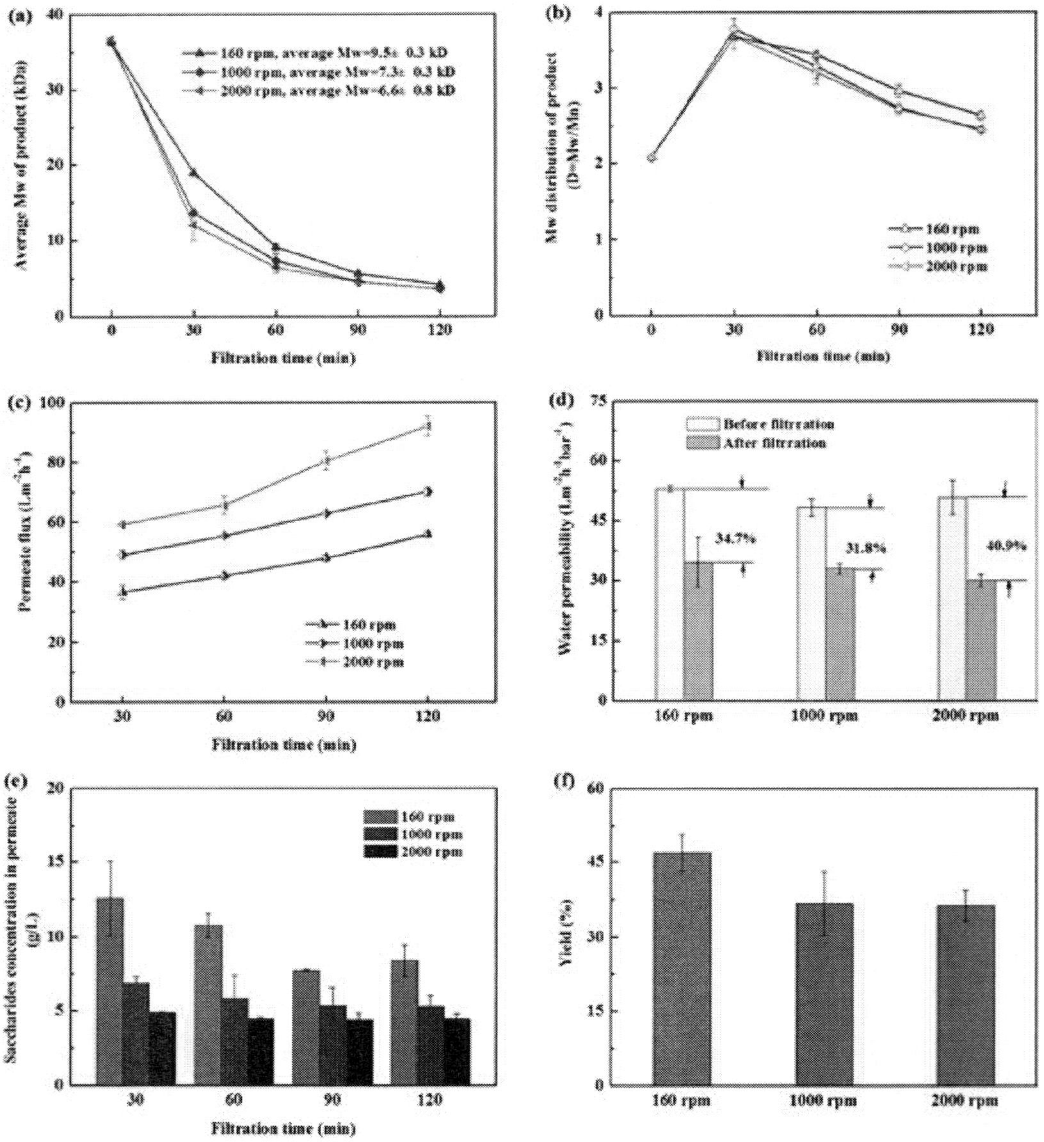


**Fig. 5.** Effect of agitation speed on oligodextran production in a PES20-EMR at 4.5 bar. (a) Average Mw, (b) Mw distribution of oligodextran molecules in permeate, (c) permeate flux with time, (d) irreversible fouling, (e) saccharides concentration in permeate and (f) yield of oligodextran in permeate.

membrane and enlarge the Mw distribution of the products in permeate. Moreover, more serious CP would cause higher irreversible fouling, which also should be controlled to maximize the transmission of target products. Therefore, effect of agitation speed (i.e. shear rate) on the EMR performance at a high TMP of 4.5 bar was evaluated in the following study.

### 3.2.2. Effect of agitation speed

As expected, higher shear rate induced by agitation reduced the average Mw of products and their Mw distribution, and also increased the permeate flux (Fig. 5a, b, c). These positive effects were attributed to a thinner CP layer at higher agitation (stronger shear-induced back diffusion). Indeed, the accumulation of large oligodextran molecules on the membrane would be remarkably alleviated, and their diffusion through the membrane would be also reduced at higher agitation speed [31]. The large oligodextrans possibly had more time to react with the enzymes in the reactor and were hydrolyzed to the target product [32], resulting in lower product Mw and distribution. Irreversible fouling only decreased slightly from 34.7% to 31.8% when the agitation enhanced from 160 to 1000 rpm (Fig. 5d), and it even aggravated to 40.9% at 2000 rpm. There is no doubt that the thinner CP layer promoted by higher shear rate would alleviate membrane fouling, however, at the same operating time, the larger permeate flux caused by higher agitation under constant TMP allowed much more solution passing through the membrane, implying that the substrate, product and enzyme could foul the membrane to a higher extent.

As above mentioned, less accumulation of oligodextran on the membrane at higher agitation decreased the diffusion of oligodextran through the membrane. On the other hand, higher permeate flux (Fig. 5c) also increased the oligodextran rejection due to the "dilution

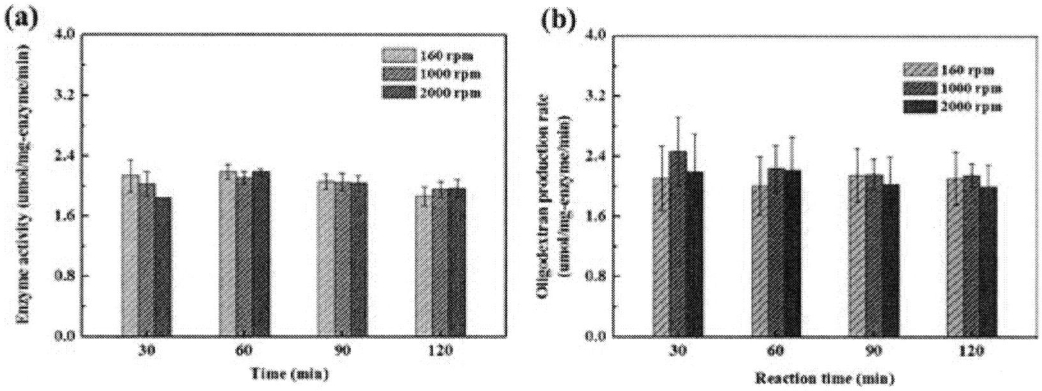


**Fig. 6.** Effects of agitation speed on (a) enzyme activity and (b) oligodextran production rate.

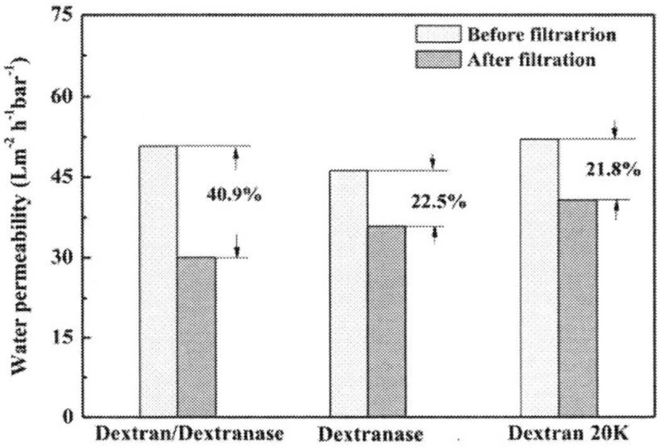


**Fig. 7.** Effect of dextran and dextranase on water permeability decline of PES20 membrane at 4.5 bar and 2000 rpm.

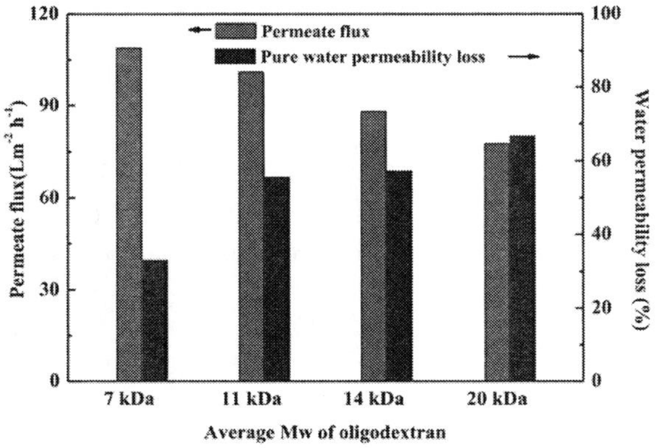


**Fig. 8.** Effect of dextran Mw on permeate flux and irreversible fouling of PES20 membrane at 4.5 bar and 2000 rpm.

effect" [18]. Consequently, the saccharides concentration in permeate decreased with increase of agitation speed (Fig. 5e). Moreover, the saccharides concentration in permeate diminished with increasing operating time probably because the substrate concentration in the EMR was decreasing under the water feeding mode while irreversible fouling was forming. The oligodextran yield was determined by the permeate flux and saccharides concentration, and the yield at 160 rpm was the highest attributed to its highest saccharides concentration in permeate (Fig. 5f). The higher permeate flux at 2000 rpm was offset by its lowest saccharides concentration, and that was why the yield did not further decline compared with that at 1000 rpm.

Besides the filtration behavior, the lower saccharides concentration in the permeate at higher agitation might be caused by the slower enzymatic hydrolysis of the substrate, as higher shear force possibly changed enzyme conformation and decreased enzyme activity [33]. To evaluate this hypothesis, the effect of the agitation speed on the enzyme activity was then investigated. Results in Fig. 6a showed that the enzyme activity at increasing incubation time (30–120 min) and different agitation speeds (160–2000 rpm) was similar, indicating that the dextranase was quite stable at high shear stress [34–36]. Moreover, the oligodextran production rate did not change significantly with time at different agitations without product removal (Fig. 6b), implying that product inhibition was negligible during the 120 min dextran hydrolysis.

Based on the above results, it could be concluded that larger TMP with high shear rate resulted in higher permeate flux and narrower Mw distribution of products, while the oligodextran yield is decreased. Higher permeate flux was preferred to timely remove the product in an EMR, while the effect of permeate flux on the oligodextran rejection should be considered [37] because the higher rejection greatly decreased the saccharides concentration in permeate, probably resulting in a yield decline. When the membrane pore size distribution was wide, controlling the CP layer was important because the unwanted dextran molecules accumulating in the CP layer would diffuse through the membrane from the large pores. Thus, selecting a suitable membrane material and controlling permeate flux, shearing velocity, and narrowing membrane pore size distribution are effective strategies to obtain high-quality oligodextran products by EMRs.

### 3.3. Fouling mechanism discussion

Membrane fouling is inevitable when operating in an EMR and has a high impact on the filtration behavior and product quality. To control and even utilize membrane fouling to our advantage, the fouling mechanisms in an EMR and their effects should be analyzed for each of the compounds occurring in the mixture that will be in contact with the membrane. It has been reported that PES membranes could be fouled by adsorption of dextran, which attaches to the membrane by Van der Waals interactions and hydrogen bonds [29,38]. Moreover, it was well known that proteins are easy to deposit in/on PES membranes by hydrophobic adsorption. Thus, the effect of substrate and enzyme on membrane irreversible fouling was systematically investigated.

#### 3.3.1. Effect of foulant type

As shown in Fig. 7, 40.9% of water permeability loss was observed after the filtration of a dextran/dextranase mixture. Two more filtrations with only dextranase or dextran (same concentration as the

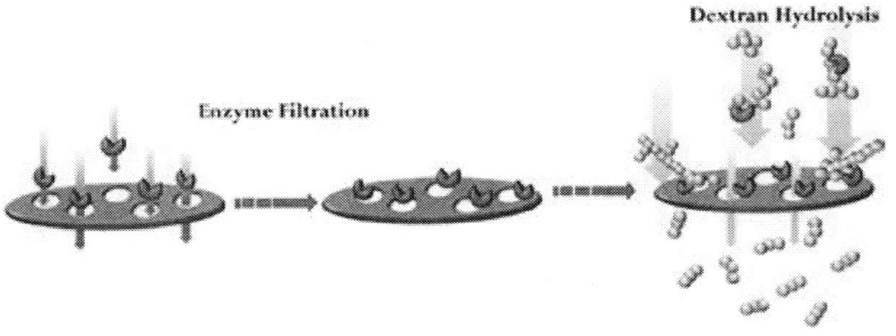


**Fig. 9.** Schematic diagram of an EMR prepared by "fouling-induced" enzyme immobilization.

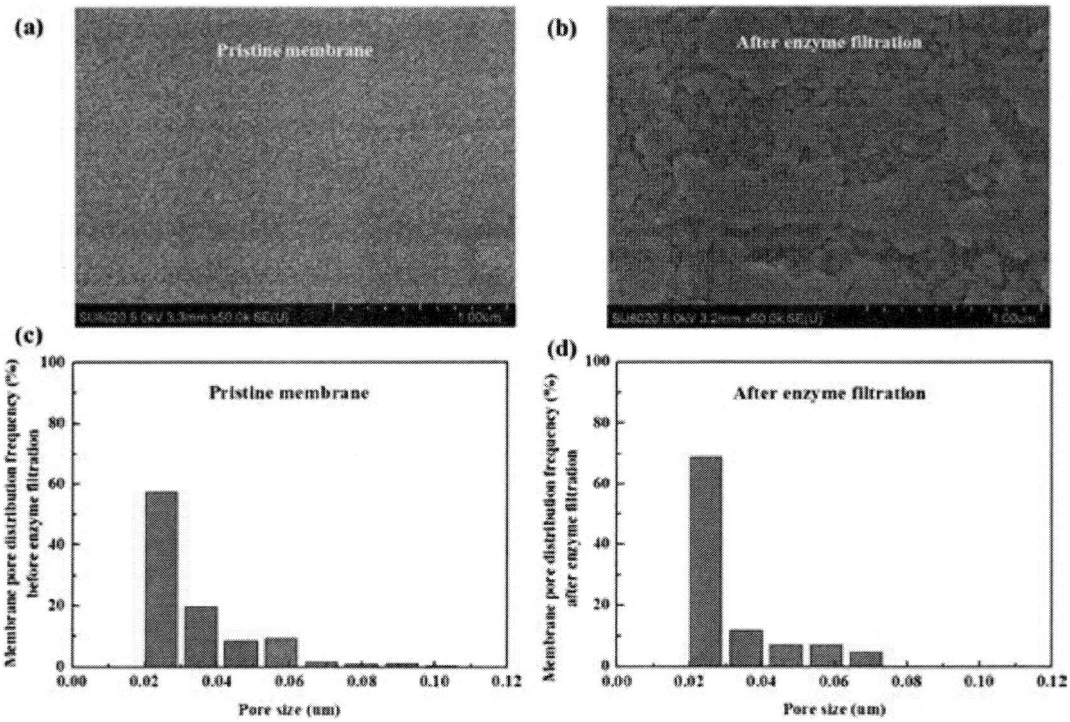


**Fig. 10.** SEM images of PES20 membrane (a) before and (b) after enzyme filtration; pore distribution frequency (c) before and (d) after enzyme filtration.

mixture) were carried out under the same operating conditions (continuously fed with water under 4.5 bar, 2000 rpm), and 22.5% and 21.8% of water permeability decline was found in the enzyme-only and dextran-only filtration systems, respectively. By these experiments it was proved that both of the enzyme and dextran molecules produced membrane irreversible fouling in an EMR.

It is worth mentioning that the concentration of dextran tested in the experiments (50 g/L) was 1000 times higher than dextranase concentration (0.05 g/L) but they caused similar membrane permeability reduction after filtration. Such result suggested that the enzymes (proteins) had larger affinity for PES membranes than dextran, which is a fact previously observed [29]. Hydrophobic and electrostatic interactions between protein and membrane have been widely investigated [39,40]. Dextranase with an isoelectric point of 3.88–4.85 was negatively charged in a neutral aqueous solution [9,35], and thus there was a negligible electrostatic attraction of dextranase with a negatively charged PES membrane. Therefore, hydrophobic rather than electrostatic interactions were probably responsible for the enzyme fouling in this case.

### 3.3.2. Effect of oligodextran Mw

It was reported that dextran Mw affected the fouling behavior on PES membranes [41]. Large dextran molecules with more contact points are expected to induce more severe fouling in/on the membrane. In this work, oligodextran molecules with different Mws were generated in the EMR during reaction, and thus it was relevant to study the effect of oligodextran Mw on fouling formation. First, DXT40K substrate was hydrolyzed by dextranase at different times, and the obtained oligodextran solutions with different average Mws (enzymes were previously deactivated) were then filtrated through the PES membrane under the same operating conditions. As shown in Fig. 8, when the average oligodextran Mw increased, the permeate flux decreased and the irreversible fouling became more serious. It was reasonable that higher rejection of larger oligodextran molecules led to thicker CP layer on the membrane, which in turn facilitated irreversible fouling formation. It was hypothesized that large dextran molecules at the beginning

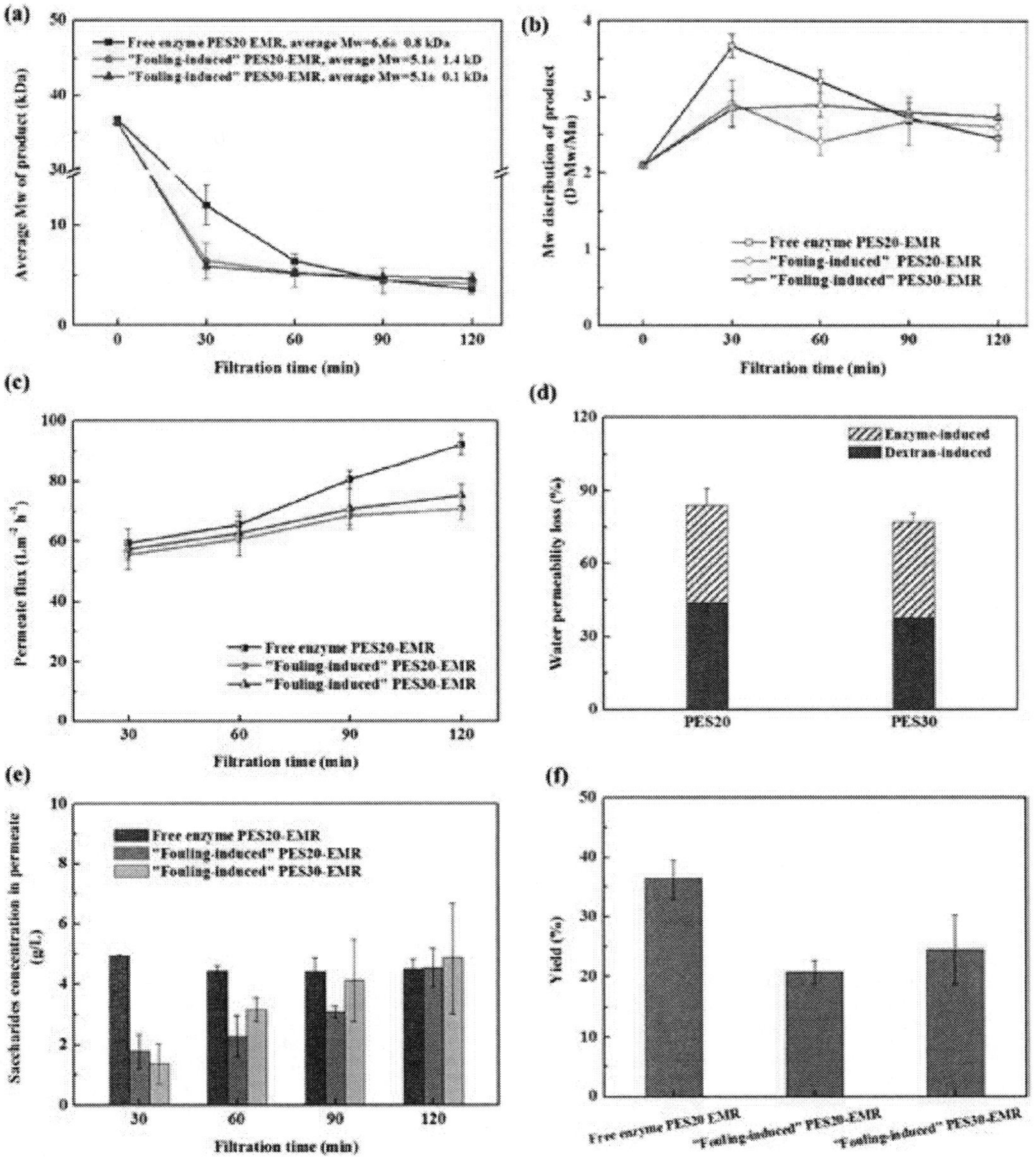


**Fig. 11.** Effect of "fouling-induced" enzyme immobilization on oligodextran production in PES-EMRs at 4.5 bar and 2000 rpm. (a) Average Mw, (b) Mw distribution of oligodextran molecules in permeate, (c) permeate flux with time, (d) irreversible fouling, (e) saccharides concentration in permeate and (f) yield of oligodextran in permeate.

of the hydrolysis dominated the fouling formation during the reaction-separation process. Interestingly, when comparing the permeability loss in Fig. 7 and Fig. 8, it was found that the irreversible fouling in the EMR (40.9%) was much lower than that observed during filtration of oligodextran with average Mw of 11–20 kDa (around 60%), implying that the large dextran molecules deposited on membrane could be hydrolyzed by the active enzyme, thus reducing the fouling evolution in an EMR.

### 3.4. *"Fouling-induced" enzyme immobilization*

Although oligodextran Mw distribution could be narrowed down by optimizing permeate flux and controlling the CP layer, it was still very wide at the beginning of operation (Figs. 4b and 5b), which was mainly ascribed to the uneven pore size distribution of the commercial membrane [42]. A self-regulating process consisting of adding certain agents capable of accessing the large pores of the membrane and might "partly" block them, leading to a reduction in membrane pore size was evaluated [43]. Theoretically, larger membrane pores would have higher local permeate flux, resulting in more serious fouling formation, while smaller pores would be less fouled [22]. Consequently, the membrane pore size distribution is expected to be narrowed during filtration, and that was possibly the reason why the oligodextran Mw distribution was decreasing from 30 min to 120 min in the cases showed in Figs. 4b and 5b. We described above how dextranase contributed more than 50% of the irreversible fouling found in the PES20-EMR membrane (Fig. 7), and it was reported that enzymes could be immobilized by the fouling formation on/in the membrane [44]. Inspired by these results, we attempted to achieve enzyme immobilization and pore size distribution narrowing by "enzyme fouling" formation, and establish a novel EMR for producing oligodextran with narrower Mw distribution (Fig. 9).

Dextranase was firstly filtrated under concentration mode to partly blocking the large pores of the membrane. PES membranes with MWCO of 20 kDa and 30 kDa were used in the experiments. As shown in Fig. 10, the surface morphology of the PES20 membrane was significantly changed after enzyme filtration, and an "enzyme fouling" layer was formed on the membrane. The membrane pore size distribution was also narrowed because some large pores disappeared and the percentage of the 20 nm pores increased from 57% to 70% (Fig. 10c and d), confirming that the enzymes were capable of blocking the larger pores during filtration.

After enzyme filtration, both the retentate and permeate of the enzyme solution were collected and mixed with the substrate, and the EMR was then operated at 4.5 bar and 2000 rpm. There was no dextranase activity detected in the permeate after enzyme filtration, however, less than 30% of protein was loaded on the membrane (detected by Bradford assay, Fig. S6), indicating that most of enzymes were still retained by the membrane and dispersed in the reactor. As seen in Fig. 11, the product Mw kept quite stable and was around 5.1 kDa for the PES20-EMR and PES30-EMR membranes with the "enzyme fouling" layer, and the Mw distribution was also obviously improved at 30–90 min, indicating that tailoring membrane pore size by "fouling-induced" enzyme immobilization was really beneficial to the product quality. Additionally the enzyme fouling treatment only reduced the permeate flux slightly as compared with the pristine EMR (Fig. 11c). The enzyme treatment, however, could not avoid the further fouling formation by oligodextrans, which resulted in a super high permeability loss (around 80%, Fig. 11d). Moreover, as shown in Fig. 11e, the reduction of pore size by "enzyme fouling" decreased the oligodextran transmission through the membrane at the beginning of the EMR operation (30–90 min), causing lower oligodextran yield for the EMR in which the "enzyme fouling" layer was induced (Fig. 11f). It is worth mentioning that with the "fouling-induced" enzyme immobilization, the oligodextran yield of the PES30-EMR was only a little higher than that of the PES20-EMR, implying that the "enzyme fouling" layer was determinant in the separation of product in this case.

Therefore, membrane fouling was not only used as a strategy for enzyme immobilization [19,20,44], but could also be useful to narrow membrane pore size and regulate its distribution, which was a promising strategy for the design of an EMR able to produce oligosaccharides and peptides with a narrower Mw distribution.

## 4. Conclusion

An EMR allowing dextran hydrolysis and in-situ product recovery with improved Mw distribution was developed in this study. When the membrane MWCO was larger than 5 kDa, the skin layer material played a more important role in the filtration as compared to the actual MWCO. A PES membrane with a nominal MWCO of 20 kDa outperformed the others since the more hydrophilic membranes (RC and PAN) had stronger interaction with dextran molecules, which led to formation of a more severe CP layer. Such a result makes us infer that the membrane material plays a particularly relevant role in the filtration of dextran (whose functional groups can interact with the main commercial membrane materials available nowadays). A more developed CP layer at higher TMPs and lower agitation rates induced more permeation of large dextran molecules through the membrane, leading to higher average Mw of products and wider Mw distribution. Moreover, serious CP seemed to increase membrane irreversible fouling, which enhanced oligodextran rejection. On the other hand, larger TMP with high agitation resulted in higher permeate flux and narrower Mw distribution of products, while the oligodextran yield might be decreased. Higher permeate flux was preferred to timely remove the product in an EMR, while the effect of permeate flux on the oligodextran rejection should be considered.

Both enzyme and dextran caused membrane irreversible fouling in an EMR, and the larger oligodextran molecules produced more fouling. Since the higher local flux around the larger membrane pores would induce more serious fouling formation, a "fouling-induced" enzyme immobilization strategy (i.e. a simple filtration of enzyme solution in advance) enabled to narrow the membrane pore size distribution of the EMR, leading to a relative constant Mw of products in permeate with narrower Mw distribution compared with the pristine EMR.

The strategy consisting of using a protein (enzyme) to foul the membrane and in turn to make more uniform the pore size distribution of the membrane has provided very promising results in this study. We think that such simple strategy could be easily implemented in other processes in which a narrower Mw distribution of products is required. A more detailed study about the influence of enzyme features e.g. size, functional groups present, etc., and operational variables like enzyme concentration and pressure during the "fouling-induced" immobilization could help generalize this strategy and extrapolate it to other processes of industrial interest.

## Acknowledgements

The authors thank the National Science Foundation of China (No. 21506229) and Youth Innovation Promotion Association (2017069) of Chinese Academy of Sciences. This work is supported by '100 Talents Program' of Chinese Academy of Sciences.

## Appendix A. Supplementary material

Supplementary data associated with this article can be found in the online version at http://dx.doi.org/10.1016/j.memsci.2018.03.062.

## References

[1] T. Heinze, T. Liebert, B. Heublein, S. Hornig, Functional Polymers Based on Dextran, Polysaccharides II, Springer, 2006, pp. 199–291.

[2] D. Kim, D.F. Day, A new process for the production of clinical dextran by mixed-culture fermentation of Lipomyces starkeyi and Leuconostoc mesenteroides, Enzyme Microb. Technol. 16 (1994) 844–848.

[3] C.E. Ioan, T. Aberle, W. Burchard, Structure properties of dextran. 2. Dilute solution, Macromolecules 33 (2000) 5730–5739.

[4] R.K. Purama, P. Goswami, A.T. Khan, A. Goyal, Structural analysis and properties of dextran produced by Leuconostoc mesenteroides NRRL B-640, Carbohydr. Polym. 76 (2009) 30–35.

[5] P. Monsan, F. Paul, D. Auriol, A. Lopez, Dextran synthesis using immobilized Leuconostoc mesenteroides dextransucrase, Methods Enzymol. 136 (1987) 239–254.

[6] D. Kim, J.F. Robyt, S.-Y. Lee, J.-H. Lee, Y.-M. Kim, Dextran molecular size and degree of branching as a function of sucrose concentration, pH, and temperature of reaction of Leuconostoc mesenteroides B-512FMCM dextransucrase, Carbohydr. Res. 338 (2003) 1183–1189.

[7] P. Barker, G. Ganetsos, N. Ajongwen, A novel approach to the production of clinical-grade dextran, J. Chem. Technol. Biotechnol. 57 (1993) 21–26.

[8] W. Gan, H. Zhang, Y. Zhang, X. Hu, Biosynthesis of oligodextrans with different Mw by synergistic catalysis of dextransucrase and dextranase, Carbohydr. Polym. 112 (2014) 387–395.

[9] A.M. Larsson, R. Andersson, J. Ståhlberg, L. Kenne, T.A. Jones, Dextranase from Penicillium minioluteum: reaction course, Cryst. Struct., Product. Complex Struct. 11 (2003) 1111–1121.

[10] S. Iqbal, R. Marchetti, A. Aman, A. Silipo, S.A.U. Qader, A. Molinaro, Enzymatic and acidic degradation of high molecular weight dextran into low molecular weight and its characterizations using novel Diffusion-ordered NMR spectroscopy, Int. J. Biol. Macromol. 103 (2017) 744–750.

[11] A.K. Goulas, J.M. Cooper, A.S. Grandison, R.A. Rastall, Synthesis of isomaltooligosaccharides and oligodextrans in a recycle membrane bioreactor by the combined use of dextransucrase and dextranase, Biotechnol. Bioeng. 88 (2004) 778–787.

[12] M. Pinelo, G. Jonsson, A.S. Meyer, Membrane technology for purification of enzymatically produced oligosaccharides: molecular and operational features affecting performance, Sep. Purif. Technol. 70 (2009) 1–11.

[13] K.C. Mountzouris, S.G. Gilmour, A.S. Grandison, R.A. Rastall, Modeling of

oligodextran production in an ultrafiltration stirred-cell membrane reactor, Enzyme Microb. Technol. 24 (1999) 75–85.

[14] K. Mountzouris, S. Gilmour, R. Rastall, Continuous production of oligodextrans via controlled hydrolysis of dextran in an enzyme membrane reactor, J. Food Sci. 67 (2002) 1767–1771.

[15] C. Torras, D. Nabarlatz, G. Vallot, D. Montané, R. Garcia-Valls, Composite polymeric membranes for process intensification: enzymatic hydrolysis of oligodextrans, Chem. Eng. J. 144 (2008) 259–266.

[16] Y. Pu, Q. Zou, L. Liu, Z. Han, X. Wang, Q. Wang, S. Chen, Clinical dextran purified by fractional ultrafiltration coupled with water washing, Carbohydr. Polym. 87 (2012) 1257–1260.

[17] M. Pinelo, V. Møller, O.A. Prado-Rubio, G. Jonsson, A.S. Meyer, Mechanisms controlling retention during ultrafiltration of charged saccharides: molecular conformation and electrostatic forces, Sep. Purif. Technol. 118 (2013) 704–709.

[18] J. Luo, S.T. Morthensen, A.S. Meyer, M. Pinelo, Filtration behavior of casein glycomacropeptide (CGMP) in an enzymatic membrane reactor: fouling control by membrane selection and threshold flux operation, J. Membr. Sci. 469 (2014) 127–139.

[19] J. Luo, A.S. Meyer, G. Jonsson, M. Pinelo, Enzyme immobilization by fouling in ultrafiltration membranes: impact of membrane configuration and type on flux behavior and biocatalytic conversion efficacy, Biochem. Eng. J. 83 (2014) 79–89.

[20] J. Luo, A.S. Meyer, G. Jonsson, M. Pinelo, Fouling-induced enzyme immobilization for membrane reactors, Bioresour. Technol. 147 (2013) 260–268.

[21] K. Zeng, J. Zhou, Z. Cui, Y. Zhou, C. Shi, X. Wang, L. Zhou, X. Ding, Z. Wang, E. Drioli, Insight into fouling behavior of poly(vinylidene fluoride) (PVDF) hollow fiber membranes caused by dextran with different pore size distributions, Chin. J. Chem. Eng. (2017).

[22] S. Ognier, C. Wisniewski, A. Grasmick, Membrane bioreactor fouling in sub-critical filtration conditions: a local critical flux concept, J. Membr. Sci. 229 (2004) 171–177.

[23] M. PleszczyŃSka, J. Szczodrak, J. Rogalski, J. Fiedurek, Hydrolysis of dextran by Penicillium notatum dextranase and identification of final digestion products, Mycol. Res. 101 (1997) 69–72.

[24] A.K. Goulas, P.G. Kapasakalidis, H.R. Sinclair, R.A. Rastall, A.S. Grandison, Purification of oligosaccharides by nanofiltration, J. Membr. Sci. 209 (2002) 321–335.

[25] X. Chen, J. Luo, B. Qi, W. Cao, Y. Wan, NOM fouling behavior during ultrafiltration: effect of membrane hydrophilicity, J. Water Process. Eng. 7 (2015) 1–10.

[26] P. Bacchin, P. Aimar, R.W. Field, Critical and sustainable fluxes: theory, experiments and applications, J. Membr. Sci. 281 (2006) 42–69.

[27] F. Meng, S. Zhang, Y. Oh, Z. Zhou, H.-S. Shin, S.-R. Chae, Fouling in membrane bioreactors: an updated review, Water Res. 114 (2017) 151–180.

[28] D.J. Miller, S. Kasemset, D.R. Paul, B.D. Freeman, Comparison of membrane fouling at constant flux and constant transmembrane pressure conditions, J. Membr. Sci. 454 (2014) 505–515.

[29] H. Susanto, S. Franzka, M. Ulbricht, Dextran fouling of polyethersulfone ultrafiltration membranes—Causes, extent and consequences, J. Membr. Sci. 296 (2007) 147–155.

[30] J. Luo, Z. Zhu, L. Ding, O. Bals, Y. Wan, M.Y. Jaffrin, E. Vorobiev, Flux behavior in clarification of chicory juice by high-shear membrane filtration: evidence for threshold flux, J. Membr. Sci. 435 (2013) 120–129.

[31] W. Zhang, L. Ding, Investigation of membrane fouling mechanisms using blocking models in the case of shear-enhanced ultrafiltration, Sep. Purif. Technol. 141 (2015) 160–169.

[32] H.E. Wray, R.C. Andrews, P.R. Bérubé, Surface shear stress and retention of emerging contaminants during ultrafiltration for drinking water treatment, Sep. Purif. Technol. 122 (2014) 183–191.

[33] S.S. Nadar, V.K. Rathod, Ultrasound assisted intensification of enzyme activity and its properties: a mini-review, World J. Microbiol. Biotechnol. 33 (2017) 170.

[34] E.R. Jiménez, Dextranase in sugar industry: a review, Sugar Tech 11 (2009) 124–134.

[35] J. Rogalski, G. Głowiak, J. Szczodrak, M. Pleszczyńska, Z. Szczodrak, A. Wiater, Purification and immobilization of dextranase, Acta Biotechnol. 18 (1998) 63–75.

[36] Y.-Q. Zhang, R.-H. Li, H.-B. Zhang, M. Wu, X.-Q. Hu, Purification, characterization, and application of a thermostable dextranase from Talaromyces pinophilus, J. Ind. Microbiol. Biotechnol. 44 (2017) 317–327.

[37] J. Luo, Y. Wan, Effect of highly concentrated salt on retention of organic solutes by nanofiltration polymeric membranes, J. Membr. Sci. 372 (2011) 145–153.

[38] H. Susanto, M. Ulbricht, Influence of ultrafiltration membrane characteristics on adsorptive fouling with dextrans, J. Membr. Sci. 266 (2005) 132–142.

[39] F. Neemann, S. Rosenberger, B. Jefferson, E.J. McAdam, Non-covalent protein–polysaccharide interactions and their influence on membrane fouling, J. Membr. Sci. 446 (2013) 310–317.

[40] Y.-N. Wang, C.Y. Tang, Fouling of nanofiltration, reverse osmosis, and ultrafiltration membranes by protein mixtures: the role of inter-foulant-species interaction, Environ. Sci. Technol. 45 (2011) 6373–6379.

[41] H. Susanto, S. Franzka, M. Ulbricht, Dextran fouling of polyethersulfone ultrafiltration membranes—Causes, extent and consequences, J. Membr. Sci. 296 (2007) 147–155.

[42] S.R. Wickramasinghe, S.E. Bower, Z. Chen, A. Mukherjee, S.M. Husson, Relating the pore size distribution of ultrafiltration membranes to dextran rejection, J. Membr. Sci. 340 (2009) 1–8.

[43] Y. Du, Y. Lv, W.-Z. Qiu, J. Wu, Z.-K. Xu, Nanofiltration membranes with narrowed pore size distribution via pore wall modification, Chem. Commun. 52 (2016) 8589–8592.

[44] J. Luo, F. Marpani, R. Brites, L. Frederiksen, A.S. Meyer, G. Jonsson, M. Pinelo, Directing filtration to optimize enzyme immobilization in reactive membranes, J. Membr. Sci. 459 (2014) 1–11.

# 522636 A Compact Double Crosslinking Technique for High Performance Solvent Resistant Nanofiltration Membrane Fabrication

**Abstract:**

Here we report a new technique to design highly stable and selective organic solvent nanofiltration (OSN) membranes with enhanced solvent permeance by manipulating the crosslinking reactions. The new technique consists of three steps: (1) membrane fabrication via non-solvent induced phase inversion by blending sulfonated polyphenylsulfone (sPPSU) and polybenzimidazole (PBI) polymers, followed by (2) crosslinking the PBI part using Î±,Î±â€²-dibromo-p-xylene (DBX) to make the membrane chemically stable, and (3) ionically crosslinking the sPPSU part with hyperbranched polyethylenimine (HPEI) to narrow down the membrane pore size without affecting the permeance significantly. Crosslinking reactions have been confirmed by FTIR and XPS analyses. The OSN performance of the double crosslinked membranes was determined by measuring the permeance of various organic solvents and the rejection rates of tetracycline (Mw = 444 $g.mol^{-1}$) as a model pharmaceutical. Depending on the testing solvents, the permeances ranged from 2 to 11.8 $L.m^{-2}h^{-1}bar^{-1}$, while the rejection rates of tetracycline varied from 67 % to 97 %. Considering the outstanding OSN performance and the great chemical stability in a wide range of solvent polarities, this novel double crosslinking technique represents a step forward in the fabrication of high performance OSN membranes.

# One step co-sintering process for low-cost fly ash based ceramic microfiltration membrane in oil-in-water emulsion treatment

*Dong Zou* [a], *Minghui Qiu* [a], *Xianfu Chen* [a], *Enrico Drioli* [b], *Yiqun Fan* [a,*]

[a] State Key Laboratory of Materials-Oriented Chemical Engineering, College of Chemical Engineering, Nanjing Tech University, Nanjing, 210009, P. R. China

[b] Institute on Membranes and Modeling of Chemical Reactors, CNR, and Department of Chemical Engineering and Materials, University of Calabria, 87030 Arcavacata di Rende, CS, Italy

*Corresponding author: Tel.: +86 25 83172277; Fax: +86 25 83172292.

E-mail: yiqunfan@njtech.edu.cn

## Abstract

Ceramic membranes with higher chemical stabilities and organic solvent resistances are indispensable for water purification. The fabrication of low-cost ceramic membranes (lower sintering consumption and cheaper membrane materials) facilitates practical water treatment applications. In this study, a novel process for the fabrication of a low-cost fly ash based microfiltration (MF) membrane via a co-sintering process was proposed. In the novel process, the MF layer was sprayed on a "green" fly ash support and co-sintered. The mullite fibers with excellent heat resistance and stability were proposed to alleviate the shrinkage difference in the co-sintering process, and improve the porosity and bending strength of the support. The raw material cost, energy consumption, and preparation period of the ceramic membrane could be significantly reduced. The prepared ceramic MF membrane had an average pore size of 100 nm with a high permeability of 450 $Lm^{-2}h^{-1}bar^{-1}$. Moreover, it exhibited a high total organic carbon (TOC) removal efficiency (> 99%) for oil-in-water (O/W) emulsion, and a high stable permeability of 165 $Lm^{-2}h^{-1}bar^{-1}$ was maintained. This technique, combining low-cost materials and the co-sintering process, can serve as a cost-effective method for the production of high-performance ceramic membranes for water purification.

**Key words:** Ceramic membrane; Low-cost; Fly ash; Co-sintering technique; Water purification

# 1 Introduction

Ceramic membranes have attracted increasing attention, due to their excellent thermal, chemical, and higher-permeable properties, thus demonstrating better performances in turbidity removal, oil-in-water (O/W) emulsion, and other important industrial applications [1-5]. Generally, ceramic membranes have an asymmetric structure consisting of a ceramic support and several layers with smaller pore sizes, and a final layer for the separation function [6]. Accordingly, the fabrication of ceramic membranes with asymmetric structures requires repeated "coating-drying-sintering" processes, which results in increased sintering energy costs, and limits the scope of their application [7]. It is therefore challenging to reduce the cost of the ceramic membranes, and to improve their competitiveness in water treatment applications [8-10]

To fabricate low-cost ceramic membranes, several researchers highlighted the co-sintering process. The co-sintering process could reduce the sintering times and shorten the preparation period, thereby reducing the cost of the membranes. For example, alumina hollow fiber membranes [11], bi-layer alumina membranes [12], alumina-zirconia ceramic membranes [13], cordierite microfiltration membranes [14], and titanium oxide ultrafiltration (UF) membranes [15] were fabricated by a co-sintering process. These studies revealed that the co-sintering process could be used for the preparation of a bi-layer membrane structure on a rigid support, by adjusting the sintering temperature, membrane thickness, and organic additives. Nevertheless, few studies have been reported on the co-sintering process for the green support and membrane layer. The support had a thickness of 2–3 mm, which was significantly larger than that of the membrane layer, and the shrinkage tension difference in the co-sintering process was proportional to the membrane thickness. Hence, it was more difficult to prepare a "support + layer" than to prepare a "layer +layer" via the co-sintering process. A one-step sintering process for ceramic microfiltration (MF) membranes was introduced to reduce the membrane cost. Moreover, two issues needed to be addressed with respect to the co-sintering process for the fabrication of a support and membrane layer: (1) the sintering temperature mismatch between the alumina support (usually requiring high sintering temperatures of 1700 ℃ and above [16]) and the ceramic MF layer (usually below 1400 ℃ ); and (2) the shrinkage difference between the green support and membrane layer.

To reduce the sintering temperature of the ceramic support, several researchers added sintering aids (boehmite sol [17], sodium dodecyl benzene sulfonate (SDBS) [18], $TiO_2$ [19], etc.) to the alumina powders, to promote the sintering process at low sintering temperatures. Another method involved the use of low-cost clays or fly ash [20] for the fabrication of the ceramic support at low sintering temperatures (typically 800–1300 ℃ ), due to the complex composition of the materials. The $Na_2O$, $K_2O$, $Fe_2O_3$ and other oxides in these materials could promote the sintering process [21]. Moreover, the cheaper clays/fly ash and lower sintering temperature could result in a significant decrease in the cost of the ceramic support. Bose et al.[22] adopted kaolinite and sawdust as cheaper materials for the fabrication of a ceramic support at 850 ℃. Kumar et al.[23] fabricated novel tubular ceramics by combining clay, kaolin, and quartz at 950 ℃, and the ceramics exhibited good acidic/basic resistances. Dong et al. [21] adopted fly ash as raw material for the fabrication of a cordierite-based ceramic support with a mean pore size of 3.6–6.4 μm at approximately 1150 ℃. The fly ash based support also ensured a sufficiently high bending strength in the application. These studies demonstrated that low-cost materials (i.e., clay or fly ash) could be used to lower the sintering temperature of the ceramic support, the cost of the ceramic membrane, and its impact on the environment [24].

Another problem in the co-sintering process was the shrinkage difference between the two layers to be sintered, which could result in severe tensile/compressive stresses [25, 26]. The tensile or compressive stresses could destroy the integrity of the membrane surface, and have a negative effect on the performance of the ceramic membranes. Addressing the mismatch problem was a prerequisite for the co-sintering process. Moreover, it was found that the mixing of the material of the bi-layer structure [27-29], the promotion of the sintering process by doping with a sintering aid [30-32], and the addition of an intermediate layer between two layers with a significantly shrinkage difference [33] were effective methods for addressing this problem. Mehranjani [34] added 5 mol% of $Fe_2O_3$ in the sub-layer and 2 mol% of $Fe_2O_3$ in the top-layer, to prevent delamination and alleviate shrinkage stress during co-sintering. Moreover, no significant delamination or microstructural defects were observed. Feng et al.[12, 13] calculated the influence of the membrane thickness on the shrinkage stress and added a sintering aid to the sub-layer, to match the shrinkage difference. The bi-layer $Al_2O_3$ MF membranes and $Al_2O_3$-$ZrO_2$ MF membrane were fabricated, and demonstrated a high potential for the treatment of O/W emulsion.

In the previous work [35], a mutual doping method was proposed to alleviate the shrinkage stress between the two layers on a rigid support, and bi-layer α-$Al_2O_3$ ceramic UF membranes were fabricated via a co-sintering process. In this study, the rigid mullite whiskers with excellent heat resistance and stability were proposed to decrease the shrinkage difference of the fly ash support in the co-sintering process. The preparation process of the fly ash based ceramic MF membranes in this study is presented in Fig. 1. The material cost and energy consumption were simultaneously reduced. In this process, certain factors with respect to the microstructures, bending strength, and separation performance were investigated in detail. The final fabricated ceramic MF membranes were used for the successful treatment of O/W emulsions.

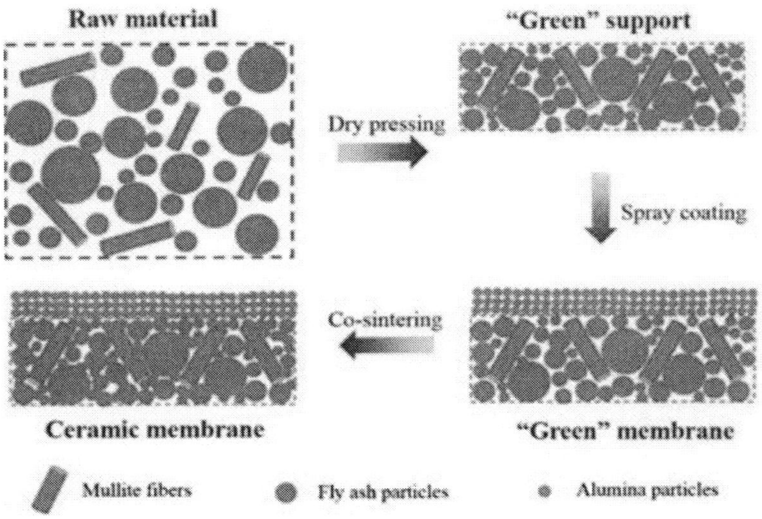


**Fig.1** The process of fabricating the fly ash based ceramic membrane.

# 2 Experimental

## 2.1 Material

The fly ash was recycled from the Gongyi electric power plant. The alumina powders with average particle sizes of 300 nm were used to prepare the MF layer. Mullite fibers (mean diameter: 11.0 μm, length: 60 μm, from Jiahe crystal fiber Co., Deqing, China) were added to the fly ash powders as a hindrance agent, to decrease the shrinkage rate of the support. Polyvinyl alcohol (PVA, Sigma) and glycerol were used as additives for the fabrication of the ceramic support. Methyl cellulose (MC, Sigma) and nitric acid (Sigma) were used as additives for the fabrication of the ceramic MF layer.

## 2.2 The preparation of ceramic membrane

### 2.2.1 The preparation of ceramic support

**Fly ash support (FS):** Fly ash, PVA, and glycerol were mixed thoroughly. The mixture was transformed into a sheet support under a pressure of 20 MPa, via a dry pressing method. The diameter and thickness of the ceramic support were approximately 30 mm and 2–3 mm. These green specimens were then sintered at different sintering temperatures (800 ℃ , 1000 ℃ , 1050 ℃ , 1100 ℃ , 1150 ℃ , and 1200 ℃ ) at a sintering rate of 2 ℃ /min.

**Mullite fibers doped Fly ash support (FMS)**: The process of the FMS was similar to that of the FS, with a certain amount of mullite fibers added to the mixture. These green specimens were dried at room temperature, for use as the support of the MF membrane.

### 2.2.2 The preparation of ceramic MF membrane

The material of the MF membrane was α-alumina. The preparation process of the alumina dispersion could be referred to our previous study [35]. The homogeneous coating suspension was sprayed onto the FMS, and the spraying time for each cycle was 5 s. After spraying different circles, these membranes were first dried at room temperature for 12 h and then dried at 110 °C for 12 h. After that, the dried membranes were sintered at 1050 °C with a hold time for 2 h (at a sintering rate of 2 °C/min). The spray apparatus was shown in Fig.S1 and some parameters were listed in the Table S1.

## 2.3 The characterization of ceramic membrane

The contents of the fly ash were measured using x-ray fluorescence (XRF) spectroscopy (Axios-Advanced, PANalytical Corporation, Netherlands). The shrinkage behaviors of the support and membrane layer were evaluated using a differential thermal dilatometer (DIL402C, Netzsch, German). The phase compositions of the fly ash were measured using x-ray diffraction (XRD, Smart Lab, Rigaku, Japan). The microstructure of the membrane or support surface was measured using field emission scanning electron microscopy (S4800, Hitachi, Japan). The roughness of the fly ash based membrane was measured using atomic force microscopy (AFM). The bending strength of the support was measured by a three-point bending test, using a spline with a length of 40 mm. More than five strips were measured for each sample. The bending strength was calculated by

$$\sigma = 3FL / 2bh^2 \tag{1}$$

where F represents the fracture force (N), L represents the length of sample (mm), b represents the width of the strip (mm), h represents the height of the strip (mm), and σ represents the bending strength (MPa). The pore size distributions of the support and membrane layer were measured using a pore size distribution analyzer (detailed information is provided in Fig. S2). The porosity of the support at different sintering temperatures was measured using the Archimedes method:

$$\varepsilon(\%) = \frac{G_3 - G_1}{G_3 - G_2} \tag{2}$$

where $G_1$ represents the weight of the dry support, $G_2$ represents the weight of the wet support measured in water, and $G_3$ represents the weight of the support that was filled with water.

## 2.4 Separation process of O/W emulsion with ceramic membrane

The O/W emulsion was home-made from oil and water with an oil concentration of 200 ppm. The fabrication process and properties of the emulsion are presented in the Fig. S3. Moreover, a cross-flow membrane separation device was used. A total organic carbon (TOC, Shimadzu ASI-5000A) device was used to detect the TOC value in the feed and permeate. The water flux (J, $Lm^{-2}h^{-1}$) and permeability (Q, $Lm^{-2}h^{-1}bar^{-1}$) were calculated using Equations (3) and (4), as follows:

$$J = \frac{V}{At} \tag{3}$$

$$Q = \frac{V}{AtP} \tag{4}$$

where $V$ represents the volume of the permeate (L), $A$ represents the effective area of the membrane surface ($m^2$), $t$ represents the filtration time in the separation process (h), and $P$ represents the trans-membrane pressure (bar). The oil removal efficiency was measured by the TOC rejection performance, and the $R_{TOC}$ was calculated using the following equation:

$$R_{TOC} = 1 - \frac{C_p}{C_F} \tag{5}$$

where $C_F$ (mg/L) and $C_P$ (mg/L) are the TOC values of the feed and permeate, respectively.

# 3 Results and discussion

## 3.1 The analysis of starting material

The chemical composition of the fly ash was obtained using XRF spectroscopy, and is presented in Table 1. It can be seen that the major components of the fly ash were $SiO_2$ (36.97 wt.%) and $Al_2O_3$ (19.79 wt.%). Moreover, other components such as $K_2O$, $Na_2O$, MgO, CaO, $P_2O_5$, $Fe_2O_3$, and $TiO_2$, were also presents in the fly ash. These oxides significantly promoted the sintering process [34], decreased the sintering temperature, and maintained a satisfactory bending strength. The carbon contents were higher than 10.68 wt.%, which could act as a natural pore forming agent.

**Table 1** Main chemical compositions (wt.%) of fly ash.

| Substance | $Al_2O_3$ | $SiO_2$ | $Fe_2O_3$ | C | CaO | $K_2O$ | $TiO_2$ | MgO | $Na_2O$ | MnO | $P_2O_5$ | Others |
|---|---|---|---|---|---|---|---|---|---|---|---|---|
| Percent (%) | 19.79 | 36.97 | 14.97 | 10.68 | 8.24 | 4.32 | 2.63 | 0.71 | 0.47 | 0.2 | 0.19 | 0.83 |

The SEM image, crystal phase, and morphology are presented in Fig. 2. As shown in Fig. 2a, the particle size of the fly ash was approximately 1–20 μm, and the particles were mainly spherical. Moreover, the spherical particles were easy to mix without ball milling. Figure 2b presents the phases of the raw fly ash with quartz, mullite, and spinel. The main phases of the fly ash were mullite ($3Al_2O_3{\cdot}2SiO_2$) and quartz ($SiO_2$) [21, 36], which is consistent with the literature. The heavy background in the XRD patterns were the impurities in the fly ash [21]. However, fly ash from different electric plants are different. It was found that a little spinel phase was present in the fly ash from the Gongyi electric plant.

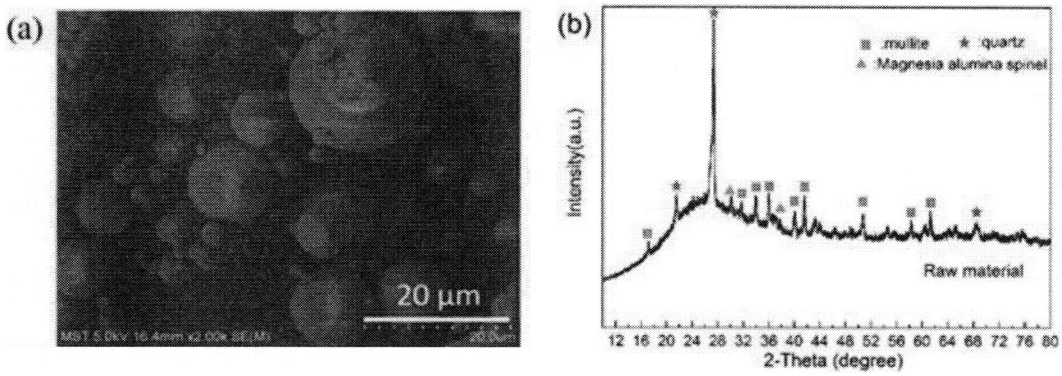


**Fig.2** The analysis of the fly ash. (a) The SEM image, (b) the phase analysis.

## 3.2 Investigation on properties of ceramic supports prepared from fly ash

To better understand the properties of the fly ash and its use as a ceramic support, the effect of the sintering temperature on the performance of the support was investigated. Figure 3 presents the effect of the temperature on the membrane surface morphology (800 °C, 1000 °C, 1050 °C, 1100 °C, 1150 °C, and 1200 °C). As shown in Fig. 3a, no significant sintering necks were observed among the fly ash particles when the sintering temperature was 800 °C. As the temperature increased, the particles began to form sintering necks, and more interconnected particles were observed. However, when the temperature was higher than 1100 °C, densification occurred. When the temperature reached 1200 °C, no pore structures were present on the surface of the support, thus indicating that the support should not be sintered at temperatures above 1100 °C. Scanning electron microscopy/energy dispersive x-ray (SEM-EDX) mapping was conducted to investigate the elemental distribution of the ceramic after sintering at 1050 °C. As shown in Fig. 3b, the main elements were Al, Si, and O. Other elements such as Fe, Ca, and Ti were dispersed homogeneously in the ceramic support, which could be used as sintering aids to increase the bending strength of the support. The results of the EDX-mapping after the sintering process were also consistent with the XRF results (the main elemental contents such as Al, O, and Si are listed in Table S2).

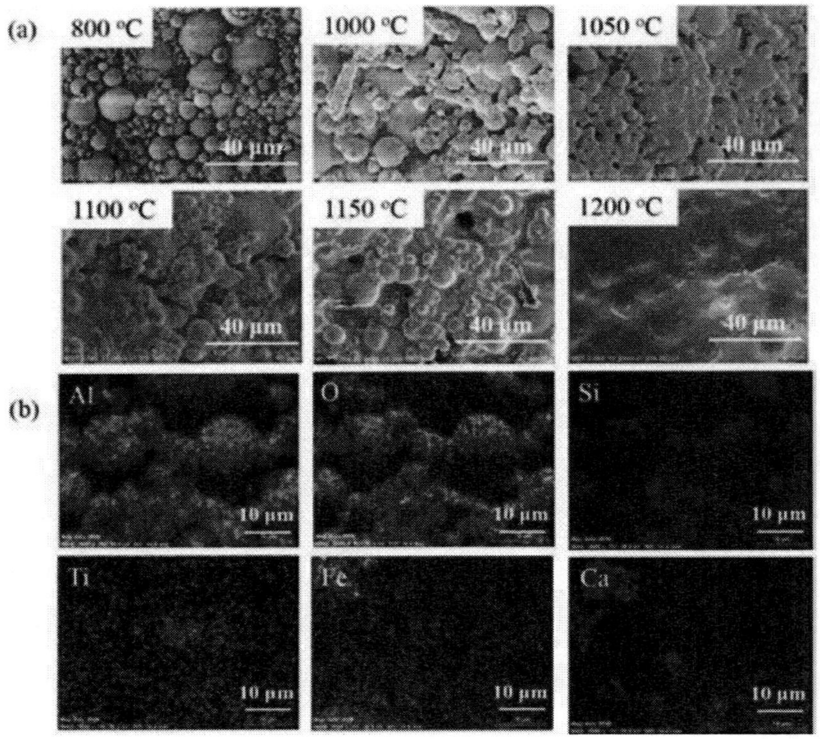


**Fig.3** The morphology of the fly ash based supports. (a) The SEM images of supports at different sintering temperatures, (b) the elemental distribution of ceramic supports from fly ash at 1050 °C.

Figure 4a presents the bending strengths of the supports at different sintering temperatures. An increase in the bending strength of the support from ~10 MPa to ~50 MPa was observed. The primary reason is that the smaller particles began to melt when the temperature increased from 1000 °C to 1200 °C. Elements such as Fe, Ca, and Ti served as sintering aids to enhance the matter transport through the liquid phase from the fly ash, and the rearrangement of larger particles under the capillary stress, during the sintering process [21]. The sintering necks could be formed to enhance the bending strength greatly. What's more, the complex composition of the fly ash promoted the phase transition of the cordierite phase and enhanced the crystallinity of the mullite (As the Fig.4b shown), which also contributed a little to the improvement of the bending strength.

Figure 4c presents the change in the water permeability under different temperatures for 2 h. It can be seen that the permeability of the support decreased from approximately 9200 to 45 $Lm^{-2}h^{-1}bar^{-1}$ when the temperature increased from 1000 °C to 1200 °C. Due to the continuous decrease in the porosity of the support (from 0.3 to 0.01), there was an increase in the filtration resistance in the test process; according to the H-P equation [37], in which the flux is proportional to the porosity (Fig. 4c). It should be noted that the pore size distribution of the support was different to that when the sintering temperature was 1150 °C. The pore size distribution of the support was wide (4-9 μm). This was because most of the fly ash particles began to melt at 1150 °C, and the fly ash liquid could completely cover the larger particles, which further promoted the growth of the fly ash particles and widened the pore size distribution. Moreover, excessive sintering could lead to a change in the original pore structure of the support. The pore size of the support at the sintering temperature from 1000 °C to 1100 °C was approximately 1–2 μm. Based on these results, the optimized sintering temperature was reported as 1050 °C.

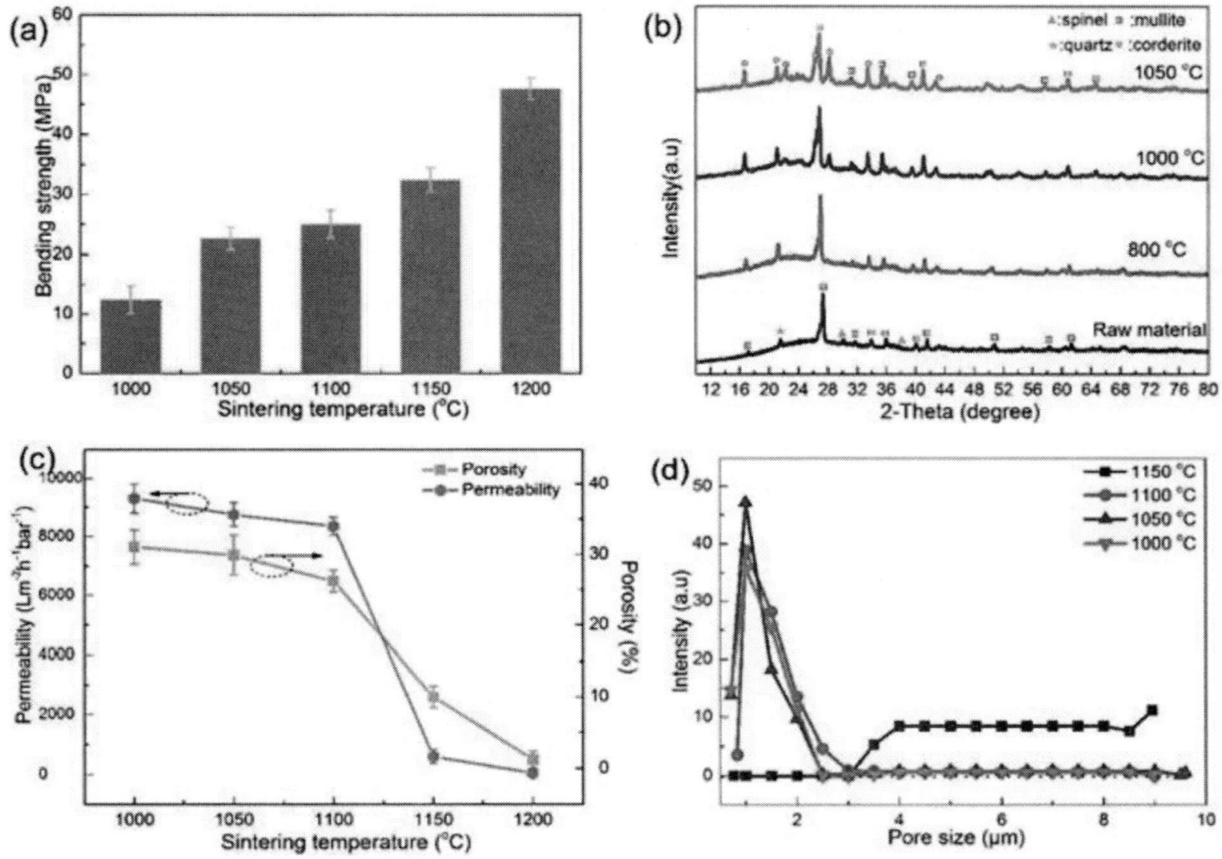


**Fig.4** The performance of the FS. (a) Bending strength at different sintering temperatures, (b) the XRD patterns of the supports at different sintering temperatures (from 0 °C to 1050 °C), (c) porosity and permeability (from 1000 °C to 1200 °C), (d) pore size distribution (from 1000 °C to 1150 °C).

## 3.3 The shrinkage difference between the support and MF membrane layer

In the previous studies [13, 15, 35, 38], the main problem to be solved with respect to the co-sintering process was the shrinkage difference between the sub-layer and top-layer. Due to the tension stress resulting from the shrinkage difference, which was proportional to the shrinkage rate [39, 40], membrane cracks occurred. Moreover, the adjustment of the shrinkage difference of the layers was a prerequisite for the co-sintering process. The commercial alumina ceramic support was more difficult to shrink than the membrane layer, given that the larger alumina particles from the support required higher sintering temperatures to form sintering necks. However, this result was inverted for the fly ash support and MF membrane layer. Figure 5a presents the shrinkage rate of the fly ash (FS) support and the MF layer. From the two curves, it can be seen that the FS support began to shrink when the temperature was 700 °C, and the shrinkage rate increased as the temperature increased. Accordingly, the shrinkage velocity of the FS support and membrane layer also increased from 700 °C to 1200 °C (Fig. S4). From the analysis of the sintering mismatch stress between the MF layer and FS support, Fig. 5b presents the classical model between the bi-layer structures in the co-sintering process. The shrinkage rate of the FS support was greater than that of the MF layer, and accordingly, the bi-layer ceramic exhibited bending deformation in the direction of the support after co-sintering. The largest sintering mismatch stress was located at the interface of the MF layer, and the stress decreased in the direction of the FS support. Given that the thickness of the support was 2–3 mm, it was more difficult to fabricate a "support + layer" than a "layer + layer" via the co-sintering process. This was because the sintering stress caused by the support was larger than that caused by the membrane layer, which was consistent with the calculations by Hsu [41] and Gurauskis [42]. The larger sintering difference could result in severe membrane cracks, and significantly decrease the performance of the ceramic MF membrane (Fig. 5c). The alleviation of the shrinkage of the support was therefore the focus of this study.

In a recent study, rigid carbon nanotubes were adopted to reduce the shrinkage rate of the alumina layer. However, the carbon nanotubes could not withstand high temperatures in an air atmosphere. In this study, low-cost mullite fibers were proposed for the reduction of the shrinkage rate, due to their excellent heat resistance and stability [43] in air atmospheres (Fig. S5 presents the stability performance at high temperatures). Figure 5a presents the shrinkage rate of the fly ash doping mullite fibers. It can be seen that the shrinkage rate was adjusted with the help of the mullite fibers. The shrinkage difference between the FMS support and the MF layer was narrower, indicating the possibility of fabricating a fly ash ceramic membrane via a co-sintering process. The SEM images of the support doping fibers after sintering at 1050 °C are presented in Fig. 5d. Moreover, the surface of the support was smooth, thus indicating that there were no negative effects of the fiber on the support surface. The possible reasons for the alleviation of the shrinkage rate by the fiber are presented in Fig. 5e. Due to the strong stability of the mullite fibers at the high temperatures, the rigid fibers served as a hindrance agent, to prevent smaller particles from excessive migration. Excessive migration could result in the severe contraction of the overall support.

Table 2 presents the investigation results of the effect of the mullite fiber doping on the porosity, bending strength, and permeability. There was a slight increase in the porosity and permeability, due to the "bridging effect" [44-46] of the mullite fibers. Moreover, it should be noted that the bending strength of the support increased from ~20 MPa to ~30 MPa, due to the formation of sintering necks between the fibers and particles; thus satisfying the requirements of industrial application [36]. Moreover, the bending strengths of the mullite fibers increased, and the bending strength of the support could also be enhanced. This was consistent with the results from the study by Zuo [47]; wherein it was reported that the addition of 0–12% of alumina fibers could increase the flexural strength of the ceramics, which could be attributed to crack deflection and frictional bridging.

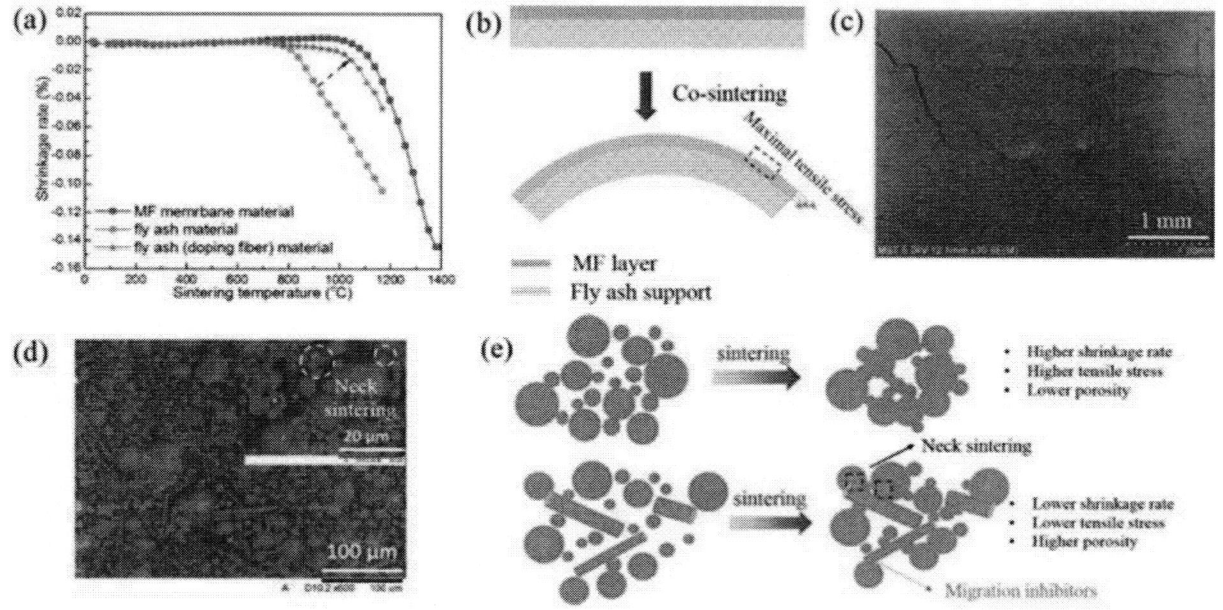


**Fig.5** The shrinkage performance of the support and the membrane material. (a) The shrinkage rate, (b) the diagram of stress in the co-sintering process, (c) membrane surface morphology after co-sintering process (the supports were fly ash without adding mullite fibers), (d) SEM images of fly ash support doping mullite fibers sintered at 1050 °C, (e) the proposed process to alleviate the shrinkage rate by doping mullite fibers.

**Table 2** The performance of the ceramic substrate with and without doping mullite fibers.

| | Bending strength (MPa) | Porosity (%) | Permeability ($Lm^{-2}h^{-1}bar^{-1}$) |
|---|---|---|---|
| Doping mullite fibers | 30.30 | 34.5 | 10500 |
| Without doping mullite fibers | 20.88 | 30.1 | 8600 |

### 3.4 Co-sintering preparation of fly ash based ceramic MF membrane

In the fabrication of ceramic MF membranes via a co-sintering process, the thickness of the MF membrane layer played a major role. Figure 6 presents the surface morphologies of the MF membranes with various spraying cycles. Figure 6a presents the support without the spraying of the alumina dispersion, which served as a control sample. From Figs. 6b and c, it can be seen that with 2 or 5 spraying cycles; the MF membrane layer could not cover the rough support, and the roughness was likely to have impact on the membrane surface, leading to an incomplete membrane surface. With 12 or 16 spraying cycles, the defects from the support were covered. However, a slight crack occurred on the membrane surface, and the cracks became severe as the number of spraying cycles increased; given that the thicker membrane layer resulted in a higher tension stress, especially in the co-sintering process [48, 49]. The higher tension stress resulted in membrane cracking [41]. Moreover, the surface of the membrane was smooth and crack-free when 8 spraying cycles were performed (Fig. 6d). The corresponding effect of the spraying times on the membrane thickness and pore size distribution are presented in the Figs. S6 and S7.

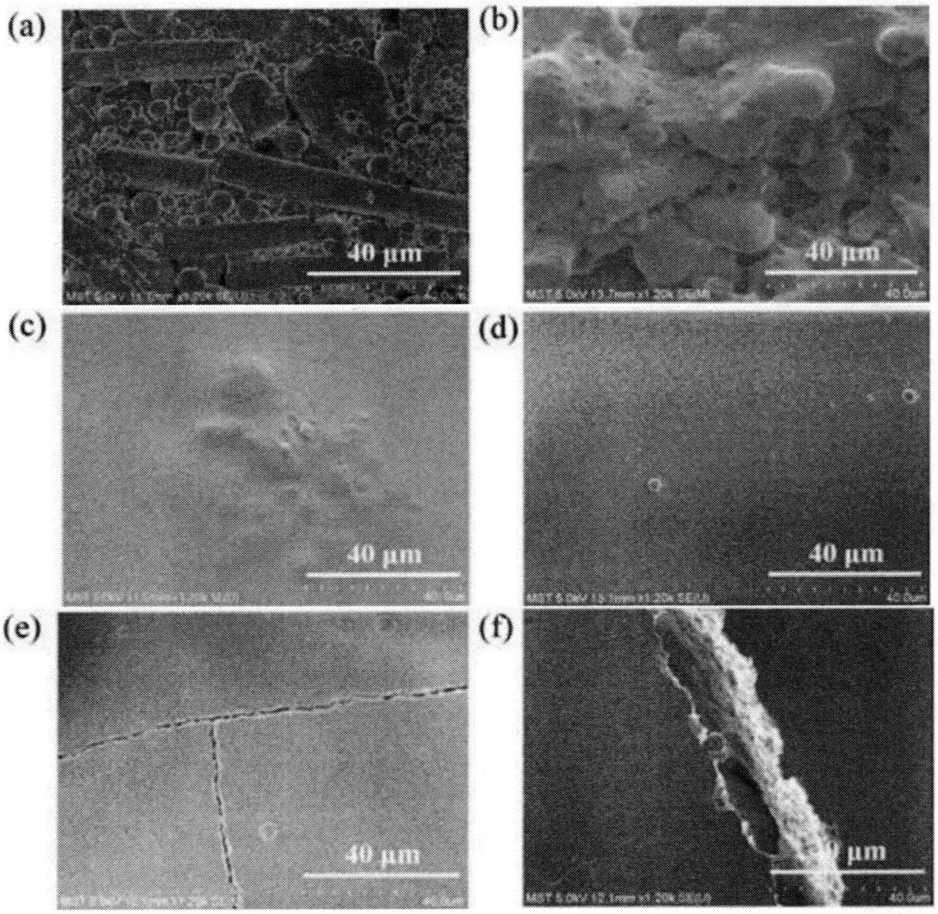


**Fig.6** The effect of the spraying cycles on the structure of the membrane. (a) 0 times, (b) 2 times, (c) 5 times, (d) 8 times, (e) 12 times, (f) 16 times.

## 3.5 The performance and emulsion treatment of MF membrane

The SEM images, roughness of the membrane surface, pore size distribution, and permeability were characterized. As shown in the high-magnification images in Fig. 7, the membrane surface was smooth and defect-free, and the particles of the membrane material were well connected. The cross-sectional structure of the MF membrane was clear, and no significant penetration phenomenon was observed (Fig. 7b). The roughness of the ceramic MF layer was evaluated according to the AFM images, with an average roughness of 25 nm (Figs. 7c, d). The AFM images indicated that the ceramic membrane had a smooth surface, which tended to minimize the deposition of pollutants in the practical application [50]. The pore size and permeability performance of the fly ash based MF membranes are presented in in Figs. 8a and b. The permeability and mean pore size of the membrane were 450 $Lm^{-2}h^{-1}bar^{-1}$ and 100 nm, respectively.

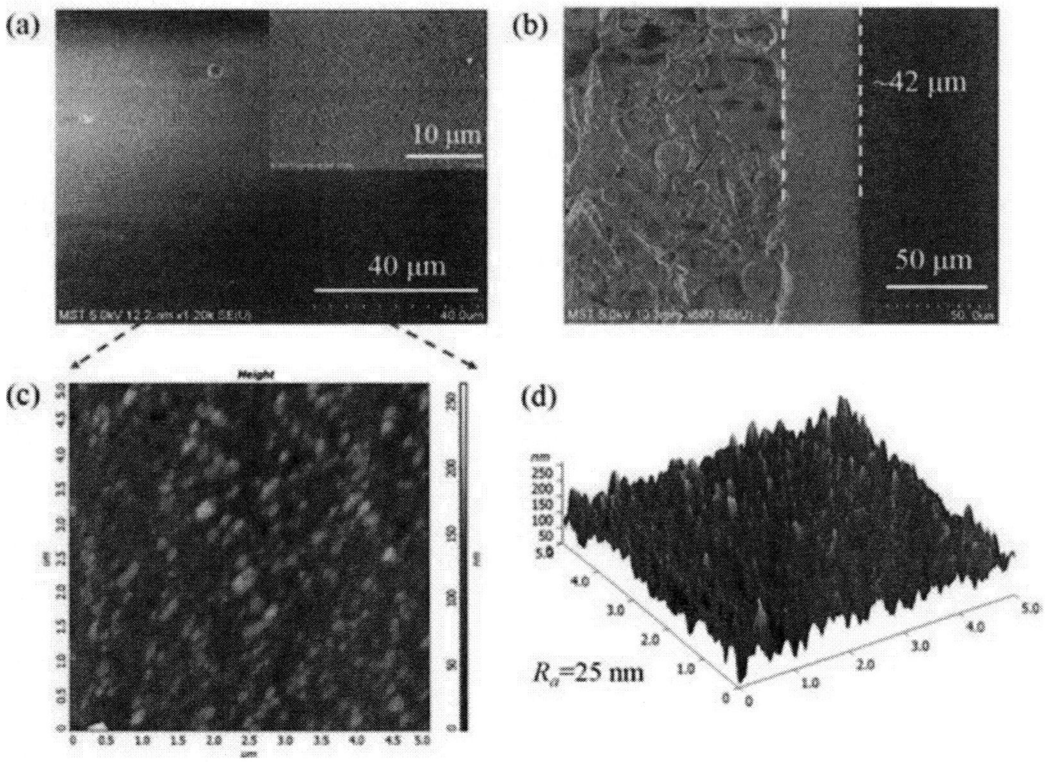


**Fig.7** The morphology of the MF membrane. (a) The SEM image of membrane surface, (b) cross section of MF membrane, (c) (d) MF membrane surface roughness.

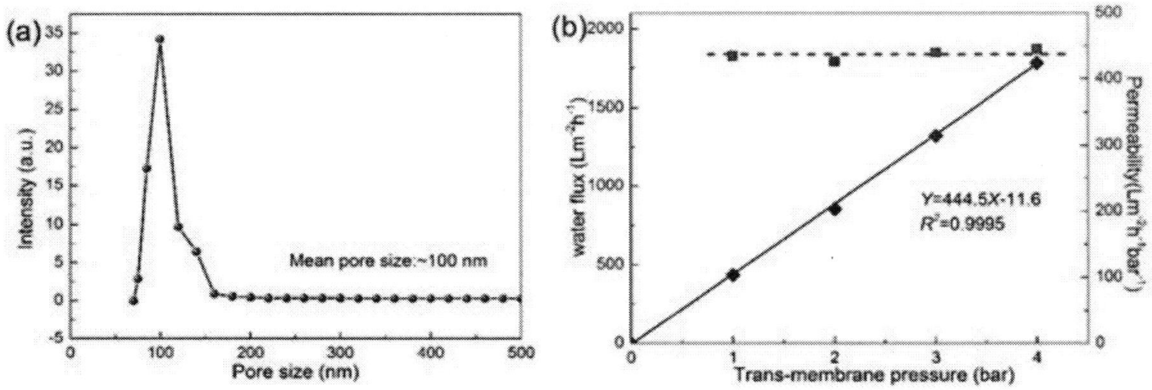


**Fig.8** The pore size distribution and permeability of fly ash based MF membrane. (a) Pore size distribution, (b) water flux and permeability.

For the analysis of the membrane performance in this study, the permeability and mean pore size were compared to those in the literature (Table 3) [23, 51-59]. The ceramic MF membrane was prepared by a conventional process, which involved the sintering of a porous alumina support at a high temperature, and then the preparation of an MF layer on the support. This process required at least two sintering steps, which required a higher energy consumption and expensive raw materials. In this study, the waste fly ash and co-sintering technique were combined to fabricate a low-cost ceramic membrane. This process greatly decreased the cost of the raw materials and the energy consumption (one sintering step required from the support to the membrane layer). The costs were qualitatively compared with those in the literature, as shown in the Fig. S8. Moreover, the fly ash based MF membrane exhibited a higher permeability than the membranes reported in previous studies, due to its smaller pore size.

The O/W emulsion test was used to evaluate the application performance of the fly ash based ceramic MF membrane. Figure 9 presents the TOC rejection and water flux of the fly ash based membrane. As shown in Fig. 9a, the TOC rejection of the membrane was higher than 99%, thus indicating the applicability of the low-cost membrane in O/W emulsion treatment (Fig. S9 presents the micrographs of the emulsion before and after treatment). However, the TOC rejection of the support ranged from 80–85% during the 300-min test. This was because the small oil drops could pass through the support pores, resulting in a lower TOC rejection. The permeability of the membrane and support are presented in Fig. 9b. The stable permeability of the MF membrane was approximately 165 $Lm^{-2}h^{-1}bar^{-1}$. The separation performance in the O/W emulsion test was also compared with those in the literature, as shown in Table 4. It can be seen that the fly ash based ceramic MF membrane demonstrated a competitive performance, due to its high TOC rejection performance and satisfactory permeability in the filtration process. To understand the fouling phenomenon of the MF layer, four classical models (Table S3) were used to analyze the process. As shown in Figs. S10 and S11, the fouling phenomenon of the MF layer was dominated by the cake layer, thus confirming that the oil could not penetrate the membrane pore; whereas the fouling model of the support was dominated by the standard blocking model. The fouling phenomenon also demonstrated that the pore size distribution of the low-cost MF membrane was uniform.

**Table 3** Comparison of the performance of the ceramic MF membrane reported in the literature.

| Author | Sintering method | Membrane material | Pore size (nm) | Permeability($Lm^{-2}h^{-1}bar^{-1}$) |
|---|---|---|---|---|
| Saja[52] | conventional | bentonite | 1300 | 520.4 |
| Song[57] | conventional | Carbon | 1000 | 450.7 |
| Lin [53] | conventional | $TiO_2$ | 380 | 1150 |
| Kumar[23] | conventional | Kaolin | 309 | 216 |
| Wang [58] | conventional | $Al_2O_3$ | 300 | 500 |
| Jedidi [56] | conventional | Fly ash | 250 | 475 |
| Yang[59] | conventional | $ZrO_2$ | 200 | 400 |
| Yu [54] | conventional | $TiO_2/Al_2O_3$ | 200 | 100 |
| Chang[55] | conventional | $Al_2O_3$ | 140 | 505 |
| Bouazizi[51] | conventional | $TiO_2$ | 78 | 35 |
| Our work | co-sintering | Fly ash/$Al_2O_3$ | 100 | 450 |

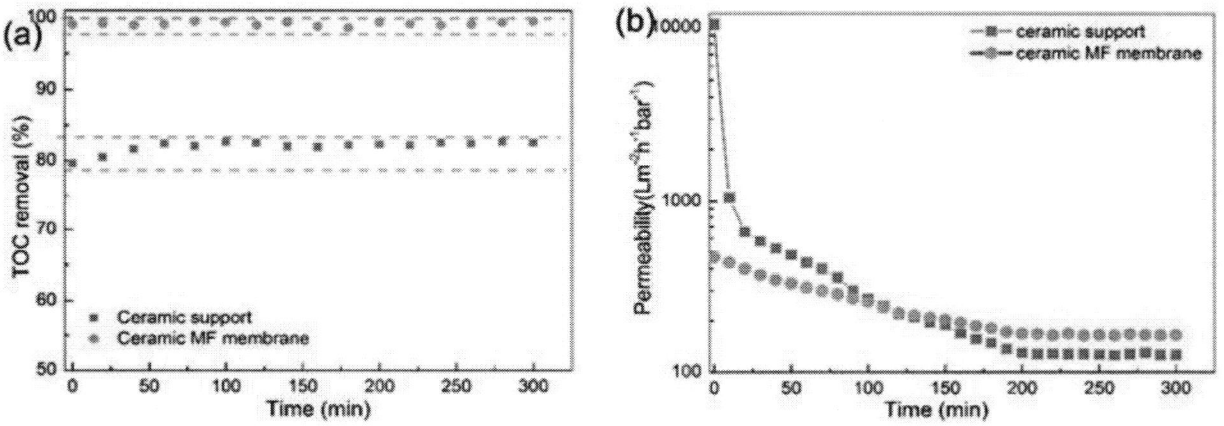


**Fig.9** The separation performance of membrane and support in treating O/W emulsion. (a) The TOC removal efficiency, (b) permeability vs. filtration time (Trans membrane pressure: 0.05 MPa, $C_f$=200 ppm).

Table 4 The comparisons of the ceramic MF membrane in treating O/W emulsion.

| | material | Pore/ oil-droplet size (μm) | Oil concentration (mg/L) | Stable permeability ($Lm^{-2}h^{-1}bar^{-1}$) | Rejection (%) |
|---|---|---|---|---|---|
| [49] | Fly ash | 0.77/2.8 | 75 | 159 | 95.3 |
| [60] | $\alpha$-$Al_2O_3$ | 0.8/4.0 | 250 | 30 | 99.9 |
| [61] | $\alpha$-$Al_2O_3$ | 0.2/20 | 141 | 230 | 97.8 |
| [62] | Carbon | 1.0/1.0 | 120 | 64.3 | 97.8 |
| [63] | $\alpha$-$Al_2O_3$ | 0.05/- | 500 | 63.9 | 98.1 |
| [24] | $TiO_2$-Mullite | 0.11/1.09 | 200 | 150 | 97 |
| This work | Fly ash/$\alpha$-$Al_2O_3$ | 0.1/0.06-10 | 200 | 165 | 99.2 |

# 4 conclusion

Fly ash and mullite fibers were used as raw materials for the fabrication of the ceramic support, and the alumina layer was then deposited on the support by spraying, using an $Al_2O_3$ suspension. After the co-sintering process, the fly ash based ceramic MF membrane was prepared by a one-step sintering process. In this process, the mullite fibers decreased the shrinkage rate of the support, which also enhanced the permeability and bending strength. After spraying the alumina dispersion (8 spraying cycles) on the support, and sintering at 1050 °C; the ceramic MF membranes could be prepared, exhibiting an average pore size of ~100 nm and permeability of ~450 $Lm^{-2}h^{-1}bar^{-1}$. The fabricated ceramic MF membrane demonstrated a higher permeability, due to its smaller pore size, when compared with those in the literature. The ceramic MF membrane was then used to separate the O/W emulsion with an oil content of 200 mg/L, and the TOC removal efficiency was approximately 99% with a stable permeability of 165 $Lm^{-2}h^{-1}bar^{-1}$. Thus, in this study, the material cost and energy consumption for the fabrication of high-quality MF ceramic membranes was successfully reduced by combining low-cost material and the co-sintering technique.

## Acknowledgement

This work is financially supported by the National Natural Science Foundation of China (91534108), National High Technical Research Program of China (2012AA03A606), and the Project of Priority Academic Program Development of Jiangsu Higher Education Institutions (PAPD). We also appreciate Dr. Han Feng (Nanjing Tech University), Mr Wu Chao (Jiangsu University), and Mr Guan Kecheng (Nanjing Tech University) for their mullite fibers and useful discussion in experimental section.

## Appendix A. Supplementary data

Supplementary data related to this article can be found online.

## Reference

[1] G. Mustafa, K. Wyns, S. Janssens, A. Buekenhoudt, V. Meynen, Evaluation of the fouling resistance of methyl grafted ceramic membranes for inorganic foulants and co-effects of organic foulants, Sep. Purif. Technol., 193 (2018) 29-37.

[2] C.-Y. Huang, C.-C. Ko, L.-H. Chen, C.-T. Huang, K.-L. Tung, Y.-C. Liao, A simple coating method to prepare superhydrophobic layers on ceramic alumina for vacuum membrane distillation, Sep. Purif. Technol., 198 (2018) 79-86.

[3] A. Oun, N. Tahri, S. Mahouche-Chergui, B. Carbonnier, S. Majumdar, S. Sarkar, G.C. Sahoo, R. Ben Amar, Tubular ultrafiltration ceramic membrane based on titania nanoparticles immobilized on macroporous clay-alumina support: Elaboration, characterization and application to dye removal, Sep. Purif. Technol., 188 (2017) 126-133.

[4] S.M. Samaei, S. Gato-Trinidad, A. Altaee, The application of pressure-driven ceramic membrane technology for the treatment of industrial wastewaters - A review, Sep. Purif. Technol., 200 (2018) 198-220.

[5] P. Geng, G. Chen, Antifouling ceramic membrane electrode modified by Magneli $Ti_4O_7$ for electro-microfiltration of humic acid, Sep. Purif. Technol., 185 (2017) 61-71.

[6] Q. Chang, Y. Wang, S. Cerneaux, J.E. Zhou, X. Zhang, X. Wang, Y. Dong, Preparation of microfiltration membrane supports using coarse alumina grains coated by nano $TiO_2$ as raw materials, J. Eur. Ceram. Soc., 34 (2014) 4355-4361.

[7] J. de Jong, N.E. Benes, G.H. Koops, M. Wessling, Towards single step production of multi-layer inorganic hollow fibers, J. Membr. Sci., 239 (2004) 265-269.

[8] L. Chen, G. Shi, J. Shen, B. Peng, B. Zhang, Y. Wang, F. Bian, J. Wang, D. Li, Z. Qian, G. Xu, G. Liu, J. Zeng, L. Zhang, Y. Yang, G. Zhou, M. Wu, W. Jin, J. Li, H. Fang, Ion sieving in graphene oxide membranes via cationic control of interlayer spacing, Nature, 550 (2017).

[9] M.A. Shannon, P.W. Bohn, M. Elimelech, J.G. Georgiadis, B.J. Mariñas, A.M. Mayes, Science and technology for water purification in the coming decades, Nature, 452 (2008) 301.

[10] X. Yang, S. Zhou, M. Li, R. Wang, Y. Zhao, Purification of cellulase fermentation broth via low cost ceramic microfiltration membranes with nanofibers-like attapulgite separation layers, Sep. Purif. Technol., 175 (2017) 435-442.

[11] A.J. Burggraaf, Fundamentals of Inorganic Membrane Science and Technology, Elsevier Science B.V. , (1996).

[12] J. Feng, Y. Fan, H. Qi, N. Xu, Co-sintering synthesis of tubular bilayer α-alumina membrane, J. Membr. Sci., 288 (2007) 20-27.

[13] J. Feng, M. Qiu, Y. Fan, N. Xu, The effect of membrane thickness on the co-sintering process of bi-layer $ZrO_2/Al_2O_3$ membrane, J. Membr. Sci., 305 (2007) 20-26.

[14] Y. Dong, B. Lin, S. Wang, K. Xie, D. Fang, X. Zhang, H. Ding, X. Liu, G. Meng, Cost-effective tubular cordierite micro-filtration membranes processed by co-sintering, J. Alloys Compd., 477 (2009) L35-L40.

[15] M. Qiu, S. Fan, Y. Cai, Y. Fan, N. Xu, Co-sintering synthesis of bi-layer titania ultrafiltration membranes with intermediate layer of sol-coated nanofibers, J. Membr. Sci., 365 (2010) 225-231.

[16] F. Bouzerara, A. Harabi, S. Achour, A. Larbot, Porous ceramic supports for membranes prepared from kaolin and doloma mixtures, J. Eur. Ceram. Soc., 26 (2006) 1663-1671.

[17] S. Ananthakumar, A.R.R. Menon, K. Prabhakaran, K.G.K. Warrier, Rheology and packing characteristics of alumina extrusion using boehmite gel as a binder, Ceram. Int., 27 (2001) 231-237.

[18] Y. Yang, F. Han, W. Xu, Y. Wang, Z. Zhong, W. Xing, Low-temperature sintering of porous silicon carbide ceramic support with SDBS as sintering aid, Ceram. Int., 43 (2017) 3377-3383.

[19] H. Qi, Y. Fan, W. Xing, L. Winnubst, Effect of $TiO_2$ doping on the characteristics of macroporous $Al_2O_3$ /$TiO_2$ membrane supports, J. Eur. Ceram. Soc., 30 (2010) 1317-1325.

[20] J. Fang, G. Qin, W. Wei, X. Zhao, Preparation and characterization of tubular supported ceramic microfiltration membranes from fly ash, Sep. Purif. Technol., 80 (2011) 585-591.

[21] Y. Dong, X. Liu, Q. Ma, G. Meng, Preparation of cordierite-based porous ceramic micro-filtration membranes using waste fly ash as the main raw materials, J. Membr. Sci., 285 (2006) 173-181.

[22] S. Bose, C. Das, Preparation and characterization of low cost tubular ceramic support membranes using sawdust as a pore-former, Mater. Lett., 110 (2013) 152-155.

[23] R.V. Kumar, A.K. Ghoshal, G. Pugazhenthi, Elaboration of novel tubular ceramic membrane from inexpensive raw materials by extrusion method and its performance in microfiltration of synthetic oily wastewater treatment, J. Membr. Sci., 490 (2015) 92-102.

[24] L. Zhu, M. Chen, Y. Dong, C.Y. Tang, A. Huang, L. Li, A low-cost mullite-titania composite ceramic hollow fiber microfiltration membrane for highly efficient separation of oil-in-water emulsion, Water Res., 90 (2016) 277-285.

[25] T.J. Garino, H.K. Bowen, Kinetics of constrained-film sintering, J. Am. Ceram. Soc., 73 (1990) 251-257.

[26] G.W. Scherer, T. Garino, Viscous sintering on a rigid substrate, J. Am. Ceram. Soc., 68 (1985) 216-220.

[27] X. Chang, C. Zhang, W. Jin, N. Xu, Match of thermal performances between the membrane and the support for supported dense mixed-conducting membranes, J. Membr. Sci., 285 (2006) 232-238.

[28] A. Rafferty, Y. Gun'ko, R. Raghavendra, An investigation of co-fired varistor-ferrite materials, J. Eur. Ceram. Soc., 24 (2004) 2005-2013.

[29] A. Rafferty, Y. Gun'ko, R. Raghavendra, An investigation of co-fired varistor-NiZn ferrite multilayers, Mater. Res. Bull., 44 (2009) 747-752.

[30] A. Chaouchi, M. Aliouat, S. Marinel, S. d'Astorg, H. Bourahla, Effects of additives on the sintering temperature and dielectric properties of $ZnTiO_3$ based ceramic, Ceram. Int., 33 (2007) 245-248.

[31] S.C. Thompson, A. Pandit, N.P. Padture, S. Suresh, Stepwise-graded $Si_3N_4$-SiC ceramics with improved wear properties, J. Am. Ceram. Soc., 85 (2002) 2059-2064.

[32] X. Zhang, C. Deces-Petit, S. Yick, M. Robertson, O. Kesler, R. Maric, D. Ghosh, A study on sintering aids for $Sm_{0.2}Ce_{0.8}O_{1.9}$ electrolyte, J. Power Sources, 162 (2006) 480-485.

[33] Q.B. Chang, L. Zhang, X.Q. Liu, D.K. Peng, G.Y. Meng, Preparation of crack-free $ZrO_2$ membrane on $Al_2O_3$ support with $ZrO_2$-$Al_2O_3$ composite intermediate layers, J. Membr. Sci., 250 (2005) 105-111.

[34] A.S. Mehranjani, D.J. Cumming, D.C. Sinclair, R.H. Rothman, Low-temperature co-sintering for fabrication of zirconia/ceria bi-layer electrolyte via tape casting using a $Fe_2O_3$ sintering aid, J. Eur. Ceram. Soc., (2017).

[35] D. Zou, M.H. Qiu, X.F. Chen, Y.Q. Fan, One-step preparation of high-performance bilayer alpha-alumina ultrafiltration membranes via co-sintering process, J. Membr. Sci., 524 (2017) 141-150.

[36] Z. Wei, J. Hou, Z. Zhu, High-aluminum fly ash recycling for fabrication of cost-effective ceramic membrane supports,

J. Alloys Compd., 683 (2016) 474-480.

[37] W. Li, W. Xing, N. Xu, Modeling of relationship between water permeability and microstructure parameters of ceramic membranes, Desalination, 192 (2006) 340-345.

[38] D. Zou, X.B. Ke, M.H. Qiu, X.F. Chen, Y.Q. Fan, Design and fabrication of whisker hybrid ceramic membranes with narrow pore size distribution and high permeability via co-sintering process, Ceram. Int., (2018). doi: 10.1016/j.ceramint.2018.08.161.

[39] B. Kellett, F.F. Lange, Stresses induced by differential sintering in powder compacts, J. Am. Ceram. Soc., 67 (1984) 369-371.

[40] P.Z. Cai, D.J. Green, G.L. Messing, Constrained densification of alumina/zirconia hybrid laminates. 1: Experimental observations of processing defects, J. Am. Ceram. Soc., 80 (2010) 1929-1939.

[41] R.T. Hsu, J.H. Jean, Key Factors Controlling camber behavior during the cofiring of bi-layer ceramic dielectric laminates, J. Am. Ceram. Soc., 88 (2005) 2429–2434.

[42] J. Gurauskis, A.J. Sanchez-Herencia, C. Baudin, Alumina-zirconia layered ceramics fabricated by stacking water processed green ceramic tapes, J. Eur. Ceram. Soc., 27 (2007) 1389-1394.

[43] F. Han, Z. Zhong, Y. Yang, W. Wei, F. Zhang, W. Xing, Y. Fan, High gas permeability of SiC porous ceramics reinforced by mullite fibers, J. Eur. Ceram. Soc., 36 (2016) 3909-3917.

[44] X. Ke, Y. Huang, T.R. Dargaville, Y. Fan, Z. Cui, H. Zhu, Modified alumina nanofiber membranes for protein separation, Sep. Purif. Technol., 120 (2013) 239-244.

[45] X. Ke, Z.F. Zheng, H.W. Liu, H.Y. Zhu, X.P. Gao, L.X. Zhang, N.P. Xu, H. Wang, H.J. Zhao, J. Shi, K.R. Ratinac, High-flux ceramic membranes with a nanomesh of metal oxide nanofibers, J. Phys. Chem. B, 112 (2008) 5000-5006.

[46] X.B. Ke, H.Y. Zhu, X.P. Gao, J.W. Liu, Z.F. Zheng, High-performance ceramic membranes with a separation layer of metal oxide nanofibers, Adv. Mater., 19 (2007) 4325-4325.

[47] F. Zuo, F. Meng, D.T. Lin, J.J. Yu, H.J. Wang, S. Xu, W.M. Guo, C. Cerecedo, V. Valcárcel, H.T. Lin, Influence of whisker-aspect-ratio on densification, microstructure and mechanical properties of $Al_2O_3$ whiskers-reinforced $CeO_2$ -stabilized $ZrO_2$ composites, J. Eur. Ceram. Soc., 38 (2017).

[48] M. Qiu, Y. Fan, N. Xu, Preparation of supported zirconia ultrafiltration membranes with the aid of polymeric additives, J. Membr. Sci., 348 (2010) 252-259.

[49] J. Fang, G. Qin, W. Wei, X. Zhao, L. Jiang, Elaboration of new ceramic membrane from spherical fly ash for microfiltration of rigid particle suspension and oil-in-water emulsion, Desalination, 311 (2013) 113-126.

[50] J. Lin, W. Ye, M.-C. Baltaru, Y.P. Tang, N.J. Bernstein, P. Gao, S. Balta, M. Vlad, A. Volodin, A. Sotto, P. Luis, A.L.

Zydney, B. Van der Bruggen, Tight ultrafiltration membranes for enhanced separation of dyes and $Na_2SO_4$ during textile wastewater treatment, J. Membr. Sci., 514 (2016) 217-228.

[51] A. Bouazizi, M. Breida, A. Karim, B. Achiou, M. Ouammou, J.I. Calvo, A. Aaddane, K. Khiat, S. Alami Younssi, Development of a new $TiO_2$ ultrafiltration membrane on flat ceramic support made from natural bentonite and micronized phosphate and applied for dye removal, Ceram. Int., 43 (2017) 1479-1487.

[52] A. Bouazizi, S. Saja, B. Achiou, M. Ouammou, J.I. Calvo, A. Aaddane, S.A. Younssi, Elaboration and characterization of a new flat ceramic MF membrane made from natural Moroccan bentonite. Application to treatment of industrial wastewater, Applied Clay Science, 132-133 (2016) 33-40.

[53] Y. Lin, D. Zou, X. Chen, M. Qiu, H. Kameyama, Y. Fan, Low temperature sintering preparation of high-permeability $TiO_2$/Ti composite membrane via facile coating method, Appl. Surf. Sci., 349 (2015) 8-16.

[54] Y. Yu, W. Hou, X. Hu, Y. Yu, L. Mi, L. Song, Superhydrophobic modification of an $Al_2O_3$ microfiltration membrane with $TiO_2$ coating and PFDS grafting, Rsc Advances, 4 (2014) 48317-48321.

[55] Q. Chang, J.-e. Zhou, Y. Wang, J. Wang, G. Meng, Hydrophilic modification of $Al_2O_3$ microfiltration membrane with nano-sized gamma-$Al_2O_3$ coating, Desalination, 262 (2010) 110-114.

[56] I. Jedidi, S. Saidi, S. Khemakhem, A. Larbot, N. Elloumi-Ammar, A. Fourati, A. Charfi, A.B. Salah, R.B. Amar, Elaboration of new ceramic microfiltration membranes from mineral coal fly ash applied to waste water treatment, J. Hazard. Mater., 172 (2009) 152-158.

[57] C. Song, T. Wang, Y. Pan, J. Qiu, Preparation of coal-based microfiltration carbon membrane and application in oily wastewater treatment, Separation & Purification Technology, 51 (2006) 80-84.

[58] P. Wang, P. Huang, N. Xu, J. Shi, Y.S. Lin, Effects of sintering on properties of alumina microfiltration membranes, J. Membr. Sci., 155 (1999) 309-314.

[59] C. Yang, G.S. Zhang, N.P. Xu, J. Shi, Preparation and application in oil-water separation of $ZrO_2$/alpha-$Al_2O_3$ MF membrane, J. Membr. Sci., 142 (1998) 235-243.

[60] J. Mueller, Y.W. Cen, R.H. Davis, Crossflow microfiltration of oily water, J. Membr. Sci., 129 (1997) 221-235.

[61] S.R.H. Abadi, M.R. Sebzari, M. Hemati, F. Rekabdar, T. Mohammadi, Ceramic membrane performance in microfiltration of oily wastewater, Desalination, 265 (2011) 222-228.

[62] C. Song, T. Wang, Y. Pan, J. Qiu, Preparation of coal-based microfiltration carbon membrane and application in oily wastewater treatment, Sep. Purif. Technol., 51 (2006) 80-84.

[63] F.L. Hua, Y.F. Tsang, Y.J. Wang, S.Y. Chan, H. Chua, S.N. Sin, Performance study of ceramic microfiltration membrane for oily wastewater treatment, Chem. Eng. J., 128 (2007) 169-175.

Food and Bioprocess Technology
https://doi.org/10.1007/s11947-018-2189-z

ORIGINAL PAPER

# Decoloration of Molasses by Ultrafiltration and Nanofiltration: Unraveling the Mechanisms of High Sucrose Retention

Shiwei Guo[1,2] · Jianquan Luo[1,2] · Qiangjian Yang[1,2] · Xiufu Qiang[1] · Shichao Feng[1] · Yinhua Wan[1,2]

Received: 22 May 2018 / Accepted: 25 September 2018


**Abstract**

Membrane technology provides a green approach to recover sucrose from cane molasses. However, the trade-off between color removal and sucrose permeation by membrane filtration debases the efficiency and limits the application. In this study, ten commercially-available and one self-made ultrafiltration (UF) and nanofiltration (NF) membranes were used for decoloration of molasses. By investigating the effect of membrane properties, molasses components, and operating parameters on the molasses filtration behaviors, the mechanisms of high sucrose retention were clarified. For polyether sulfone (PES) membrane, the high retention of pigments and sucrose was mainly caused by the serious irreversible fouling as an additional selective layer. Polyamide (PA) NF membranes showed high antifouling ability, and a suitable pore size (~ 500 Da) was important to achieve high color removal and sucrose permeation. Although regenerated cellulose (RC) membranes exhibited excellent antifouling performance to molasses, the severe membrane swelling induced by salts and high temperature (60 °C) limited their application. Reducing sugar in molasses produced negligible effect on the sucrose retention, while inorganic salts resulted in pore swelling and solute dehydration due to "salting-out" effect, thus improving the sucrose permeation. However, such positive effect was weakened or eliminated by the pigments in molasses, and irreversible fouling became more serious in the presence of the molasses salts, especially for PES membranes. Based on the underlying mechanisms, the pore swelling and sucrose dehydration effects were "reappeared" by simply removing the non-polar pigments via macroporous resin adsorption, thus decreasing sucrose retention. High temperature could accelerate the sucrose permeation, but it also attenuated concentration polarization layer, thus intensifying the sucrose retention increase induced by high permeate flux. If a loose NF membrane was selected for decoloration of molasses, minimizing fouling formation, reappearing salt effect on membrane and sucrose, operating at high temperature and low permeate flux could minimize sucrose retention.

**Keywords** Salt-induced pore swelling · Membrane fouling · Loose nanofiltration · Tight ultrafiltration · Decolorization

## Introduction

Bioproduct raw materials (e.g., fermentation broths, plant extracts, and enzymatic hydrolysate) are important sources of many necessary substances (drug, health care products, condiments, food additives, cosmetics etc.) for human. To obtain high-quality target products, decoloration of raw materials is often required. Therefore, a number of technologies were used to remove or recover the pigments such as adsorption by resin (Blanc et al. 2017), chitosan (Xu et al. 2016) and active carbon (Bernal et al. 2016), chemical degradation (Nguyen et al. 2017), biodegradation (Hai et al. 2011), extraction (Zhou et al. 2017), and so on. However, the requirements of adsorbent pretreatment/regeneration and adding extra chemicals for these methods would induce some environment and safety problems. Membrane technology, especially ultrafiltration (UF) and nanofiltration (NF), provides a green method for separation and purification thanks to its simple operation, small footprint, no chemicals addition, and mild operation conditions (Russin et al. 2011; Ambrosi et al. 2014), which has been applied in color removal of

✉ Jianquan Luo
jqluo@ipe.ac.cn

✉ Yinhua Wan
yhwan@ipe.ac.cn

1 State Key Laboratory of Biochemical Engineering, Institute of Process Engineering, Chinese Academy of Sciences, Beijing 100190, People's Republic of China

2 University of Chinese Academy of Sciences, Beijing 100049, People's Republic of China

Published online: 06 October 2018

Springer

fermentation broth (Mai et al. 2018), fermentation wastewater (Liu et al. 2013), soy sauce (Miyagi et al. 2013), palm oil production effluent (Subramaniam et al. 2017), and sugarcane juice (Luo et al. 2016).

Cane molasses is the main byproduct in the sugar production process, which mainly contains sucrose, glucose, fructose, salts, pigments, and some colloids (Sharma et al. 2016). Thereinto, the salts and pigments are the main components that impede sugar recovery from molasses. That is why, the cane molasses is mainly utilized as the substrate for fermentation (Sguarezi et al. 2009; Colla et al. 2010). However, due to the presence of excessive inorganic salts and recalcitrant pigments, the fermentation using molasses as substrate results in a large amount of hypersaline and high strength wastewater (Satyawali and Balakrishnan 2008), which is difficult to deal with and also devalues molasses in fermentation industry. Therefore, decoloration of cane molasses followed by desalination for sugar recovery is imperative and meaningful from both environmental and economic perspectives.

There were few reports regarding decoloration of cane molasses by UF and NF. Qi et al. (2017) preliminarily investigated the effect of pretreatment methods on color removal from cane molasses by tight UF, and the color removal efficiency was only around 60%. For other sugar-related raw materials, Hamachi et al. (2003) obtained a maximum color removal of 59% when using 1000 Da NF membrane to treat raw brown sugar, and further, decoloration with adsorbents was still required. Even with 500 Da NF membrane, the color removal was only 76–80% for green syrup (Djurić et al. 2004; Gyura et al. 2005). While in a pilot-scale sugarcane juice refining test, we found that 96% of color in fresh sugarcane juice could be removed by an integrated membrane system (Luo et al. 2016). This could be explained by the difference in pigment composition between fresh juice and molasses, and many small pigments were produced during the subsequent chemical clarification and evaporation processes. Theoretically, selecting a NF membrane with suitable pore size can achieve high enough decoloration efficiency for molasses, but the sucrose retention by such NF membrane may become a bottleneck of this application. Djurić et al. (2004) found that during the decoloration of syrup by a polyamide NF membrane (500 Da), the average sucrose retention was 45%, whereas Goulas et al. (2002) reported that the sucrose retention reached 72% when using a cellulose membrane (1000 Da) in the separation of oligosaccharides. In our pilot-scale test for sugarcane juice decoloration, a diafiltration operation was required to recover the sucrose accumulated in the retentate of the tight UF, and we found that the sucrose retention was 25% first and then even increased to 55% during the diafiltration process (Luo et al. 2016). High sucrose retention not only leads to sugar loss but also causes severe concentration polarization (i.e., high osmotic pressure) consuming more energy (Luo et al. 2018). Moreover, the sucrose recovery by diafiltration needs a large amount of water and also decreases productivity. Therefore, breaking the trade-off between color removal and sucrose permeation during UF and NF of molasses is meaningful for the sucrose recovery from molasses in industry. To realize this, elucidating the underlying mechanisms of high sucrose retention by membranes is essential. However, to the best of our knowledge, there was no systematic investigation regarding sucrose retention behavior during molasses decoloration by membranes, and most of the studies focusing on sucrose separation were only conducted with simple model solution (Moreno-Vilet et al. 2014; Wang et al. 2002; Zhao et al. 2013).

In present study, the decoloration of cane molasses by ten commercially available and one self-made UF and NF membranes was carried out under constant volume diafiltration model. We attempted to figure out the mechanism of high sucrose retention by systematically studying the effect of molasses components (i.e., reducing sugar, salts, pigments) and operating parameters (i.e., temperature, permeate flux, and shear rate) on the solute retention, transmembrane pressure (TMP), and irreversible fouling (i.e., permeability loss) of the selected membranes. Diafiltration and concentration polarization models were used to analyze the experimental results for further understanding the filtration behaviors. Based on the underlying mechanisms, a simple solution for minimizing sucrose retention was proposed. The outcome of this work is not only helpful to minimize sucrose retention during cane molasses decoloration by membrane filtration, but also provides guidance for novel membrane development and process design in industrial production.

## Experiment

### Chemicals and Membranes

Sugarcane molasses was kindly provided by a local sugar mill in Zhanjiang, Guangdong Province, China. The raw molasses was firstly diluted using deionized water (mass ratio, molasses/water = 1:2), then adjusted to pH 7.0 ± 0.2 by directly dissolving solid sodium hydroxide, finally centrifuged at 5000*g* for 15 min to remove most of the colloids and suspended matter. The obtained supernatant was used for membrane filtration tests. Model solutions were prepared by, respectively, dissolving sugar (sucrose and glucose) and salts (NaCl, KCl, $CaCl_2$, $MgSO_4$) in deionized water according to the molasses composition (sucrose: 152 g $L^{-1}$; reducing sugar: 44 g $L^{-1}$; salts: 54 mS $cm^{-1}$). The salt composition and concentration in diluted cane molasses detected by ion chromatography are shown in Table 1. Desalted molasses was prepared through diafiltrating the molasses by NF5 membrane to remove the salts, and the lost sugar was replenished by adding sucrose and reducing sugar according to the molasses composition. Ten times water was

**Table 1** Ion types and concentrations in diluted and centrifuged cane molasses

| Ions | $Na^+$ | $K^+$ | $Ca^{2+}$ | $Mg^{2+}$ | $Cl^-$ | $SO_4^{2-}$ |
|---|---|---|---|---|---|---|
| Concentration (mg $L^{-1}$) | 1262 ± 88 | 13,488 ± 211 | 1059 ± 85 | 1423 ± 58 | 6016 ± 114 | 7408 ± 159 |

used during diafiltration to remove 91.3% of conductivity in the diluted molasses while 99.3% of color was kept.

A non-polar macroporous resin (HPD722-resin) was purchased from Cangzhou Bon Adsorber Technology Co., Ltd. China. UF and NF membranes used in this work are all listed in Table 2 with their main characteristics, on the basis of the manufacturers' information. Specifically, contact angle of the NF5 and UP005 membranes was 27° and 74°, respectively. The NF5 membrane was fabricated through interfacial polymerization process with 0.1% piperazine (water phase) and 0.08% trimesoyl chloride (organic phase) with certain additive in piperazine solution. The contact time of PES membrane substrate in water phase and organic phase is 500 s and 10 s, respectively. The membranes were tailored into proper shape for the dead-end filtration setup. The membrane disks were firstly dipped in ethanol for about 5 s to remove manufacturing residues from the membrane surface and then soaked in deionized water for at least 12 h prior to use.

## Experimental Setup and Procedures

The decoloration of molasses was carried out through a constant volume diafiltration process using a dead-end filtration cell with magnetic-driven stirring. The detailed information of this bench-scale setup has been depicted in literature (Luo et al. 2010). The volume of the cell was 13 mL, with a membrane disk having diameter of 24 mm and effective area of $4.52 \times 10^{-4}$ $m^2$. The ultrapure water was pumped into the cell directly through a high-performance positive displacement pump (P-500, Pharmacia, Sweden) which could set the permeate flux. The temperature was controlled by water bath and the stirring speed was adjusted using magnetic agitator. The transmembrane pressure (TMP) was measured by a pressure sensor connected to the computer. The permeate was collected for certain volume in vials for analysis. Membrane permeability was measured with deionized water before and after the filtration process to calculate irreversible fouling for each test. The same flux was used for each kind of membrane and the pressure was kept in the certain range of 5–10 bar for all the membranes. The detailed testing conditions are listed in the captions of figures for corresponding results (for example, $V/V_f = 1$ means that the permeate volume was the same as the feed volume, 13 mL), and all the experiments were repeated at least three times to ensure the reliability of data.

## Analytical and Characterization Methods

pH value of diluted molasses was measured by SevenCompact S210 pH meter (Mettler Toledo, Switzerland) fitted with a temperature compensator. The conductivity was obtained using a conductivity meter (DDS-307A, Precision & Scientific Instrument Co., China). The color of diluted molasses was determined spectrophotometrically at 420 nm using a UV-9000S spectrophotometer (Shanghai Yuanxi Instruments Co., Ltd., China). Sucrose content was determined through Roe colorimetric method (Luo et al. 2018). Reducing sugar concentration was qualified by the method of Miller with dinitrosalicylic acid reagent (Miller 1959). Ion chromatography (761 compact IC, Metrohm, Switzerland) was used to measure the concentrations

**Table 2** Main characteristics of the UF and NF membranes tested

| Membrane type | Manufacturer | Materials | Molecular weight cut-off (MWCO) | Water permeability ($Lm^{-2}$ $h^{-1}$ $bar^{-1}$)[a] |
|---|---|---|---|---|
| NP010-1K | Microdyn-Nadir | Polyether sulfone | 1000 Da | 14.0 |
| PE-2K | Meiyi | Polyether sulfone | 2000 Da | 8.5 |
| PES-3K | Ande | Polyether sulfone | 3000 Da | 9.7 |
| UP005P-5K | Microdyn-Nadir | Polyether sulfone | 5000 Da | 21.4 |
| RC-1K | Millipore | Regenerated cellulose | 1000 Da | 1.9 |
| RC-3K | Millipore | Regenerated cellulose | 3000 Da | 5.0 |
| RC-5K | Millipore | Regenerated cellulose | 5000 Da | 5.5 |
| NTR7450 | Nitto-Denko | Sulfonated polyether sulfone | 600–800 Da | 4.6 |
| PES1 | Ande | Polyether sulfone | 800–1000 Da | 8.7 |
| NF5 | Self-made | Polyamide | 340–460 Da[a] | 14.8 |
| TS40 | Meiyi | Polyamide | 400 | 5.2 |

[a] Own measurement at 25 °C and 5 bar

of cations and anions in molasses, as a reference for the preparation of model solution.

The hydrophilicity of membrane was tested using an optical contact angle goniometer (OCA20, Data Physics Instruments Co. Germany). Each test was performed by dropping a water droplet of 2 μL on the membrane surface and then immediately recording the contact angle value. Each measurement was repeated at least three times. Membrane surface charge property was measured by an Electro Kinetic Analyzer (SuRPASS, Anton Paar Co., Austria) in a background electrolyte of 1 mM potassium chloride at various pH and 25 ± 1 °C. The pH was adjusted with 0.5 M hydrochloric acid and 0.5 M potassium hydroxide. Surface morphology of the initial and used membranes was observed with a JSM-6700F Field Emission Scanning Electron Microscopy (SEM) (JEOL Co., Japan). Prior to SEM analysis, the membrane samples were sputtered with a thin layer of gold using a JFC1600 sputter coater (JEOL Co., Japan). The molecular weight cut-off (MWCO) of the membranes was measured using several representative sugar (glucose-180 Da, sucrose-342 Da, raffinose-504 Da, α-cyclodextrin-973 Da, and β-cyclodextrin-1135 Da). The single sugar solution (1000 ppm) was filtrated using the dead-end setup at 25 °C, 1200 rpm, and 40 $Lm^{-2}$ $h^{-1}$. The MWCO value of each membrane is defined as the molecular weight of a neutral solute that is rejected to 90%. The sugar concentration in the feed and permeate was determined by total organic carbon (TOC) analyzer (Shimadzu TOC-L).

## Data Processing and Modeling

The observed retention ($R_{obs}$) of pigment, sucrose, reducing sugar, and conductivity by membranes was calculated as follows:

$$R_{obs} = \left(1 - \frac{C_p}{\frac{1}{2}(C_f + C_r)}\right) \times 100\% \quad (1)$$

where the $C_p$, $C_f$, $C_r$ are the concentration (sugar), absorbance (pigment), or conductivity (salt) of the permeate, feed, and retentate, respectively.

Permeate flux ($J_p$) was calculated as follows:

$$J_p = \frac{V_p}{t \cdot A_m} \quad (2)$$

where $V_p$ is the permeate volume, $A_m$ is the effective membrane area, t is the filtration time.

Membrane permeability, $L_p$, was calculated as follows:

$$L_p = \frac{J_p}{TMP} \quad (3)$$

Membrane irreversible fouling (IF) was defined as percentage of permeability loss after filtration:

$$IF(\%) = \frac{L_{p,b} - L_{p,a}}{L_{p,b}} \times 100\% \quad (4)$$

where $L_{p,b}$ and $L_{p,a}$ are the $L_p$ before and after filtration, respectively.

In constant volume diafiltration, the loss of solute from the feed is equal to the mass in the permeate, that is,

$$V_f dC = -C_p dV \quad (5)$$

where the $V_f$ is the volume of feed solution, $C_p$ is the concentration of solutes in the permeate.

Substituting Eq. (1) into Eq. (5), and then integrating it at $V = 0$, $C = C_f$, gives,

$$\frac{C_r}{C_f} = \exp\left((R_{obs} - 1)\frac{V}{V_f}\right) \quad (6)$$

According to film theory (Bowen and Jenner 1995), the solute concentration at membrane surface ($C_{i,m}$) was calculated by following equation.

$$J_V = k \ln\left(\frac{C_{i,m} - C_{i,p}}{C_{i,f} - C_{i,p}}\right) \quad (7)$$

where $J_v$ (m $s^{-1}$) is the volumetric flux per unit area of membrane, $C_{i,f}$ and $C_{i,p}$ are the solute concentrations in feed and permeate, respectively, k is the mass transfer coefficient, which could be calculated by an empirical relationship (Opong and Zydney 1991):

$$\frac{kr}{D_{I,\infty}} = \psi\left(\frac{\omega r^2}{v}\right)^{0.567}\left(\frac{v}{D_{I,\infty}}\right)^{0.33} \quad (8)$$

where r (m) is the radius of the cell, ω (rad $s^{-1}$) is the angular velocity, v (Pa·s) is the kinematic viscosity, $D_{I,\infty}$ ($m^2$ $s^{-1}$) is the diffusion coefficient of solute at infinite dilution, Ψ is the constant which depends on the module geometry and could be experimentally determined. Here, the value is 0.335 for the filtration cell used in this work (Bowen et al. 1997).

The osmotic pressure (Π) and its difference (ΔΠ) across the membrane can be calculated according to the following equations:

$$\Pi = nRT = \frac{C}{M}RT \quad (9)$$

$$\Delta\Pi = \Pi_{int} - \Pi_p \quad (10)$$

where *n* is the molar number, *R* is the gas constant (8.314 J $mol^{-1}$ $K^{-1}$), *T* (K) is the Kelvin temperature, *C* (g $L^{-1}$) is the mass concentration of solute, *M* is the molar mass of solute (g $mol^{-1}$). $\Pi_{int}$ and $\Pi_p$ (Pa) are the osmotic pressures at the membrane surface and in the permeate, respectively.

## Results and Discussion

### Effect of Membrane Material and Pore Size

The membrane filtration performance is mainly determined by its intrinsic properties (surface material, pore size, etc.) (Schmidt et al. 2016). Thus, selecting a suitable membrane may simultaneously realize high color removal and sucrose permeation for the molasses refining. On the other hand, by comparing the filtration behaviors of molasses with different membranes, the solute features in the molasses can be understood. As shown in Fig. 1, for each kind of solute in the molasses, their retentions for different membranes followed the same variation trend, that is, a membrane with higher color removal would definitely have larger retentions of sugar and salts. By comparing the solute retention for each membrane, it was found that they were in the same order: pigment > sucrose > reducing sugar > salts, which was obviously in agreement with their molecular size. In fact, sucrose and reducing sugar are neutral while pigments and salts are charged. The domination of molecule size in their retention behaviors implied that the charge effect between membrane and solutes was greatly weakened by the high salt concentration in molasses (Table 1) (Tanninen et al. 2006).

Unexpectedly, the PE-2K and UP005 membranes showed much higher sucrose retention than other UF membranes even with smaller MWCO. This could be explained by pore size narrowing caused by the severe membrane fouling for these two polyether sulfone (PES) membranes, and this reason was supported by their highest irreversible fouling as shown in Fig. 1 (the PE-2K membrane also displayed the highest TMP). For the NTR7450 and TS40 NF membranes, their

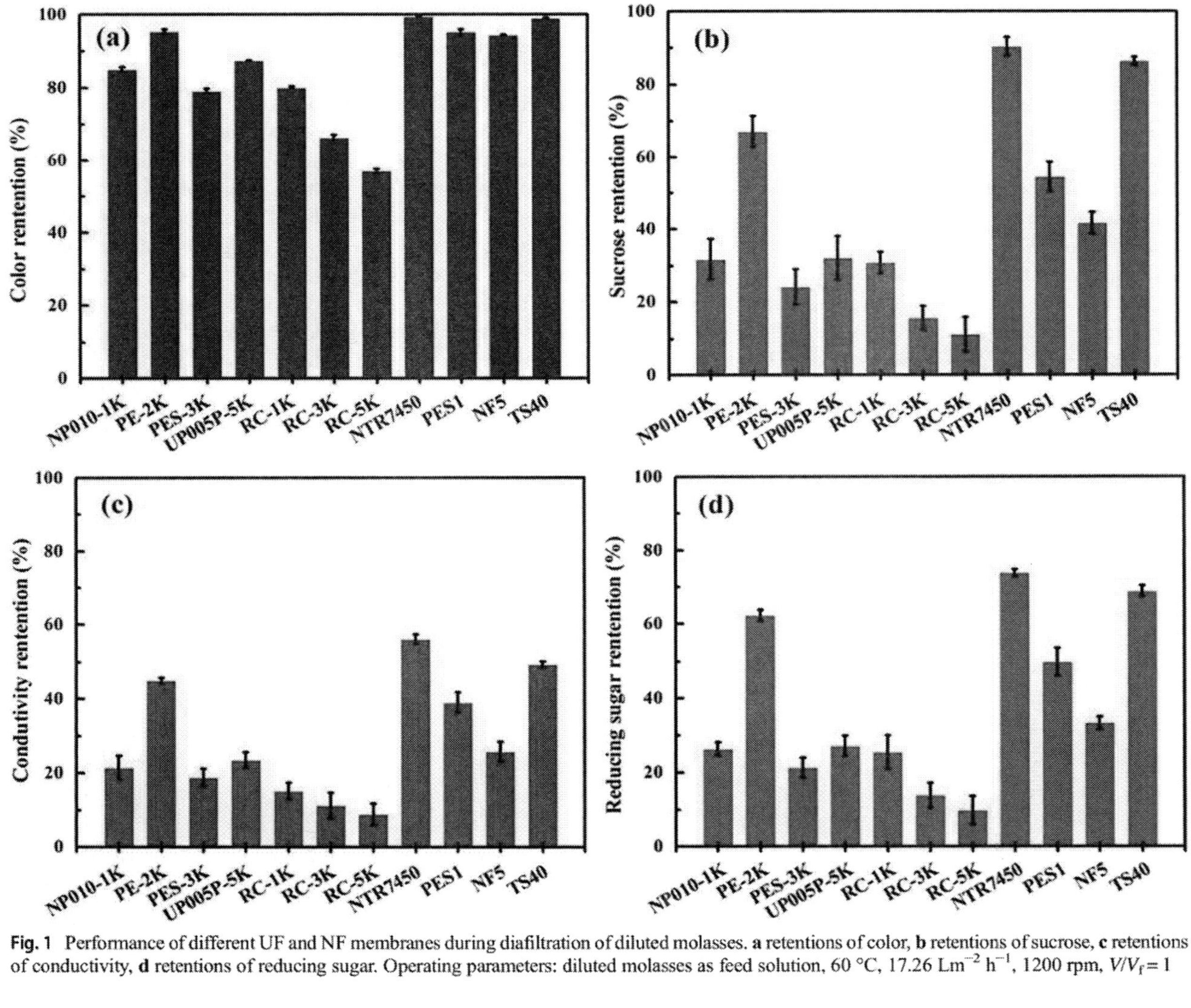


Fig. 1 Performance of different UF and NF membranes during diafiltration of diluted molasses. **a** retentions of color, **b** retentions of sucrose, **c** retentions of conductivity, **d** retentions of reducing sugar. Operating parameters: diluted molasses as feed solution, 60 °C, 17.26 $Lm^{-2}h^{-1}$, 1200 rpm, $V/V_f = 1$

sucrose retention was highest, more than 80%, but their retention mechanisms were different. The MWCO of the NTR7450 membrane was 600–800 Da, which was much larger than the molecular weight of sucrose (342 Da), whereas its irreversible fouling was 67.3% (Fig. 2b), indicating that the fouling layer might be responsible for its high sucrose retention. For the TS40 membrane, its MWCO was only 400 Da and the irreversible fouling was 11.6%, implying that its own separation layer dominated the sucrose retention.

As seen in Fig. 2a, the TMP during the diafiltration of diluted molasses was mainly governed by the membrane MWCO and irreversible fouling, where the membrane with smaller pore size and severer permeability loss showed higher TMP because the filtration resistance consisted of both membrane and fouling layer resistances. Although the initial permeability of regenerated cellulose (RC) membranes was smaller than that of other membranes (except for NTR7450 and TS40), the TMP of these three RC membranes was lowest, which might be caused by their super high hydrophilicity. It was worth mentioning that the permeability of all the RC membranes unexpectedly increased after filtration (Fig. 2b). This was probably due to its excellent anti-fouling ability as well as membrane swelling effect induced by the high salt concentration and high temperature (Manttari et al. 2002; Luo and Wan 2013). Moreover, it was found in Fig. 2b that the irreversible fouling was mainly determined by membrane materials, where the PES membranes showed the most serious fouling (67–90%), and the permeability loss was only 7–12% for the polyamide (PA) membranes.

For the decoloration of molasses by membrane filtration, the sucrose retention should be as low as possible on the premise of high color removal (> 85%). Thus, the UP005 and NF5 membranes were selected for further investigations. Although the MWCO difference of these two membranes is huge (5000 vs. < 500 Da), their TMP during diafiltration was quite close (10–12 bar) as shown in Fig. 2a. As the irreversible fouling of UP005 was 80.7%, its high color removal might be mainly owning to the dense fouling layer. Figure 3 shows the zeta potential and MWCO measurements of these two membranes, confirming that both NF5 and UP005 membranes are negatively charged at pH 7 and the MWCO of NF5 is much smaller than that of UP005. Moreover, the contact angle results indicate that the PA NF5 membrane is much more hydrophilic than the PES UP005 membrane. Therefore, the much severer membrane fouling of UP005 was mainly caused by hydrophobic adsorption and cake layer formation.

As seen in Fig. 3b, the MWCO of the NF5 membrane at 60 °C was 460 Da, higher than that measured at 25 °C (335 Da), indicating that the high operating temperature was required and membrane fouling should be minimized in order to ensure a high permeation of sucrose (342 Da). However, for UP005, the formation of fouling layer in/on the pristine membrane was the key for the high color removal. Besides the interaction of membrane and solutes (i.e., membrane fouling), the interaction among the solutes in molasses would also affect sucrose retention, which were systematically investigated in the next section.

## Effect of Molasses Components

Molasses is composed of many substances, mainly including sucrose, reducing sugar, salts, pigments, and colloids. In order to figure out how the solutes influence the sucrose retention, three model solutions (made of single sucrose, sucrose/reducing sugar, sucrose/reducing sugar/salts, respectively), desalted molasses (equivalent to sucrose/reducing sugar/pigment solution), and original diluted molasses were used as feed solutions for the diafiltration tests.

Figure 4a shows the sucrose retention of five kinds of feed solutions by the NF5 and UP005 membranes. It can be seen that the reducing sugar had little influence on the sucrose retention for both membranes. However, with addition of

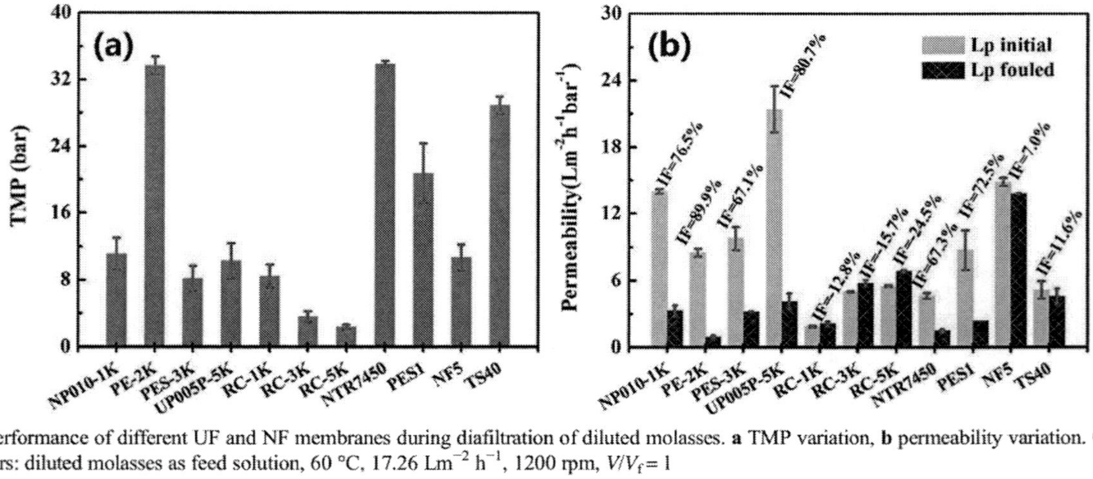


**Fig. 2** Performance of different UF and NF membranes during diafiltration of diluted molasses. **a** TMP variation, **b** permeability variation. Operating parameters: diluted molasses as feed solution, 60 °C, 17.26 Lm$^{-2}$ h$^{-1}$, 1200 rpm, $V/V_f = 1$

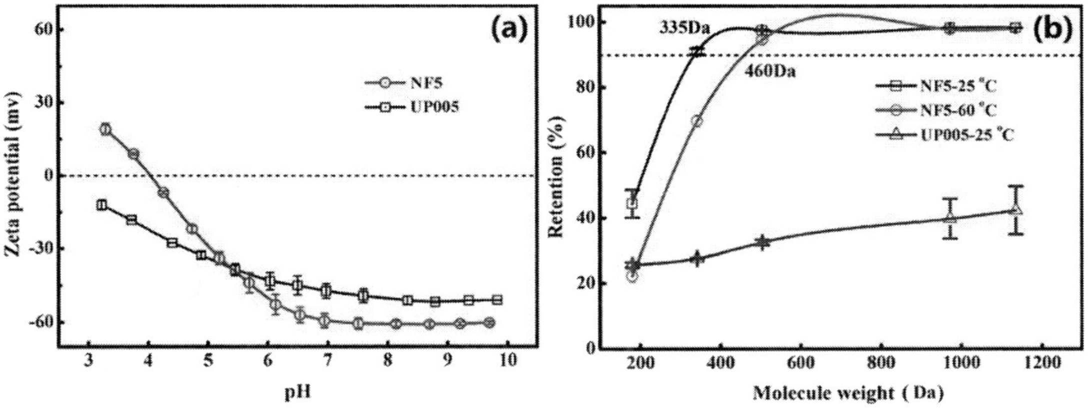


**Fig. 3** Characterization of UP005 and NF5 membranes: **a** zeta potential, **b** molecular weight cut off at different temperatures. Operating parameters: initial testing solute concentration = 1 g/L, diafiltration mode, 39.8 $Lm^{-2}$ $h^{-1}$, 1200 rpm, $V/V_f = 0.5$

salts, the sucrose retention by both membranes greatly decreased, from 46 to 21% and from 14 to 8% for NF5 and UP005, respectively. This can be mainly explained by the salt-induced pore swelling and sucrose dehydration (Boy et al. 2012; Bargeman et al. 2014; Escoda et al. 2010). In detail, the effective membrane pore size became larger (the hydration layer on pore wall became thinner) and the hydration radius of sucrose molecule became smaller (the hydration layer on solute surface became thinner), both leading to a lower sucrose retention. For desalted molasses, the sucrose retention of the NF5 kept almost the same as the sucrose/reducing sugar solution, while the UP005 exhibited an obvious increase of sucrose retention to 21% (Fig. 4a), indicating that the pigments in molasses resulted in membrane fouling as an additional separation layer for the UP005, but not for the NF5 (Fig. 4b). Interestingly, for diluted molasses, the sucrose retention of the NF5 remained the same as that of using desalted molasses or sucrose/reducing sugar solution as feed. That is, the salt-induced pore swelling and sucrose dehydration disappeared in the presence of pigments in molasses. The similar permeability loss of the NF5 for diluted molasses and sucrose/reducing sugar solution also confirmed this deduction (Fig. 4b). This means that the existence of pigments in molasses can counteract such salt effect. Two possible mechanisms were proposed as shown in Fig. 5. First, the pigments might preferentially interact with salts, and the "salting-out" effect would act on the pigments rather than the membrane and sucrose (competitive dehydration); thus, the hydration layers on membrane pore wall and sucrose molecule surface were kept, resulting in a recovery of the effective pore size and sucrose molecule size; second, the "salting-out" effect still happened on the membrane and sucrose, but the pigments would form fouling layer on the membrane with thinner hydration layer (less hydrophilic, more serious foulant adsorption), and such fouling layer offset the pore enlarging and sucrose shrinking induced by "salting-out" effect (fouling neutralizing). As seen in Fig. 4, the sucrose retention by the UP005 for diluted molasses was much higher than that for desalted molasses due to severer membrane fouling caused by the pigments in the presence of salts, implying that the second mechanism was more suitable to explain the results obtained by the UP005 membrane.

Generally, the salt-induced pore swelling effect was more obvious for the NF5 membrane with high hydrophilicity, and

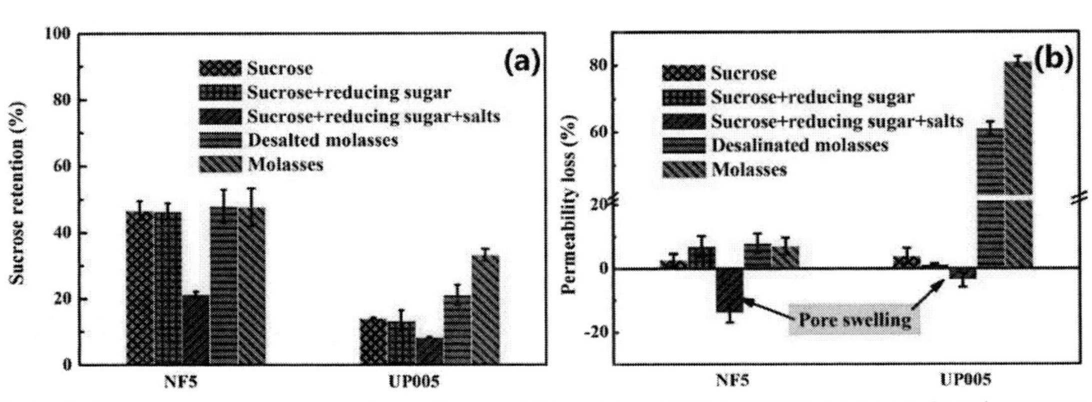


**Fig. 4** Effect of feed components on **a** sucrose retention and **b** permeability variation of NF5 and UP005 membranes. Operating parameters: 60 °C, 17.26 $Lm^{-2}$ $h^{-1}$, 1200 rpm, $V/V_f = 1$

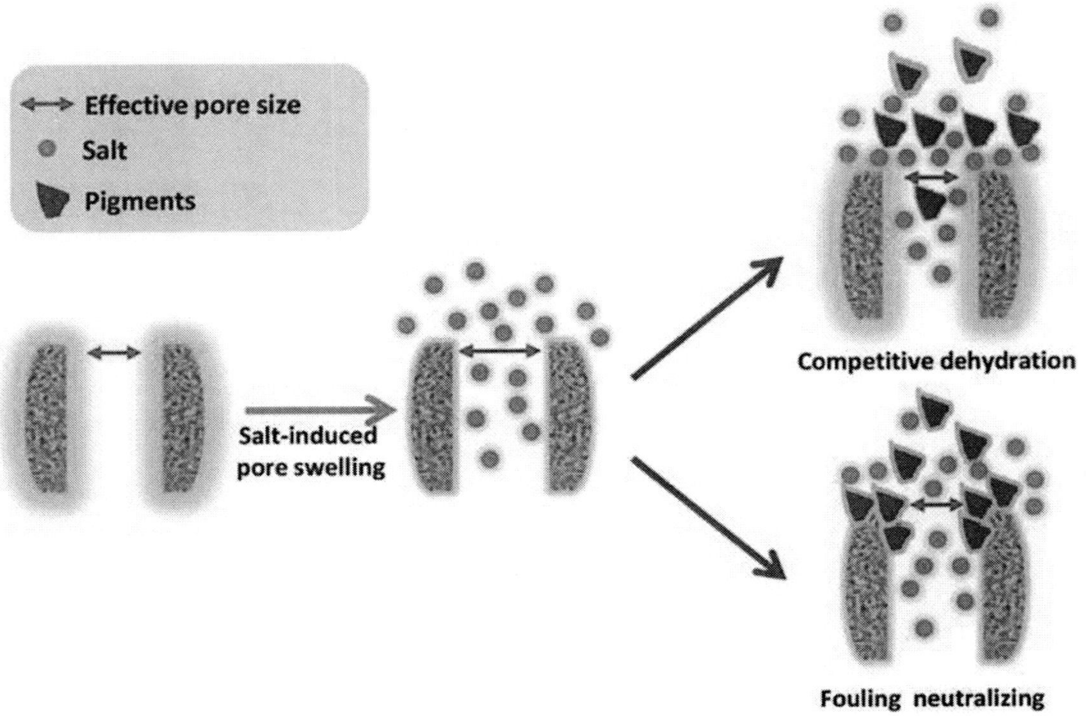


**Fig. 5** Schematic diagram for proposed mechanisms of "salting-out" effect and foulant-induced pore recovery

this NF membrane with thicker hydration layer had stronger anti-fouling ability but was more sensitive to "salting-out" effect compared with the UP005 membrane. Moreover, as shown in Fig. 6, the TMP variation for different feed solutions was also distinct for these two membranes. For the NF5, the TMP decreased a lot with addition of salts in model solution because of the pore swelling and sucrose dehydration. However, compared with the sugar solution without salts, the presence of pigments produced little effect on TMP due to its strong anti-fouling performance. But for the UP005, the TMP for the three model solutions was almost the same because of negligible osmotic pressure difference across this loose membrane, and the filtration of desalted molasses showed higher TMP due to the membrane fouling caused by pigments. When using diluted molasses as feed, both membranes showed a significant increase in TMP attributed to more serious concentration polarization and higher sucrose retention. Besides, the TMP elevation of the UP005 was also induced by severer fouling formation (Fig. 4b). The above explanations were also supported by the SEM surface images of the pristine and used membranes in Fig. 7, where surface fouling was observed on the UP005 membrane while the

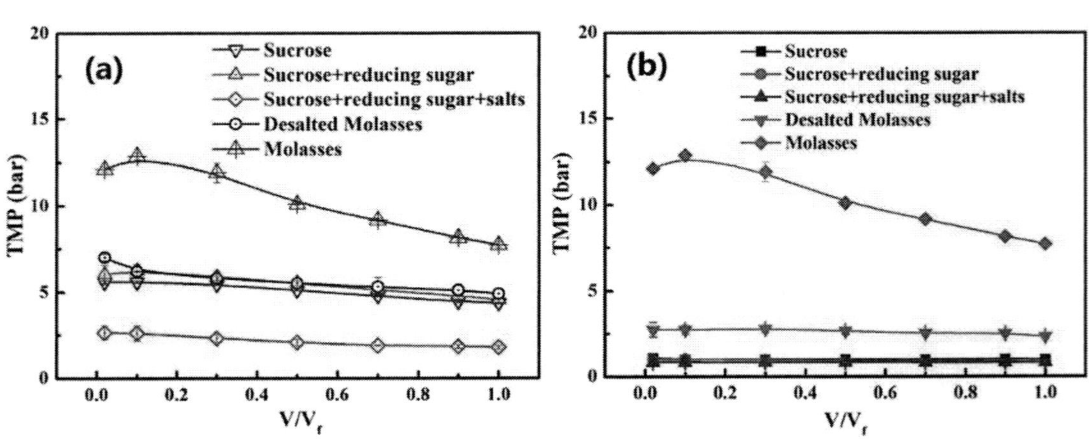


**Fig. 6** TMP variation during diafiltration of **a** NF5 and **b** UP005 membranes using feeds with different components. Operating parameters: 60 °C, 17.26 $Lm^{-2}$ $h^{-1}$, 1200 rpm, $V/V_f = 1$

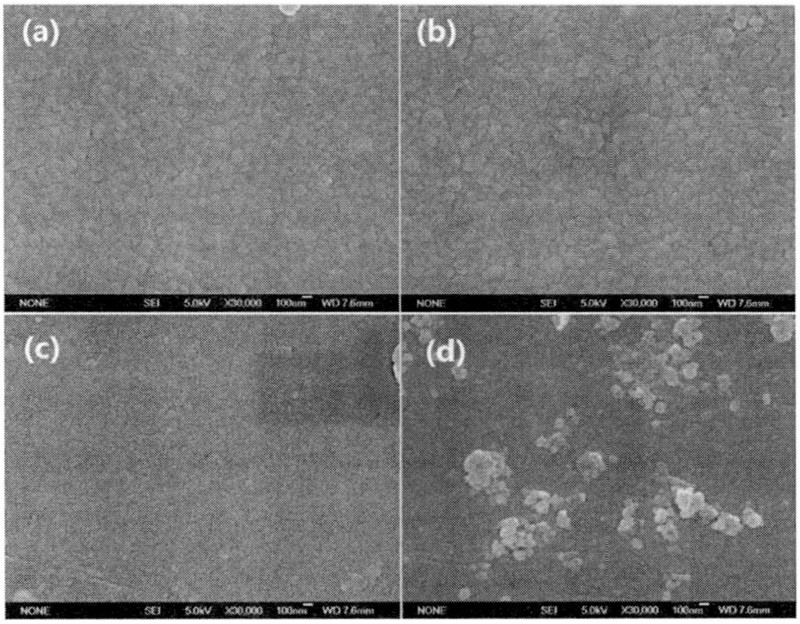


**Fig. 7** SEM surface images of **a** pristine and **b** used NF5 membranes; **c** pristine and **d** used UP005 membranes

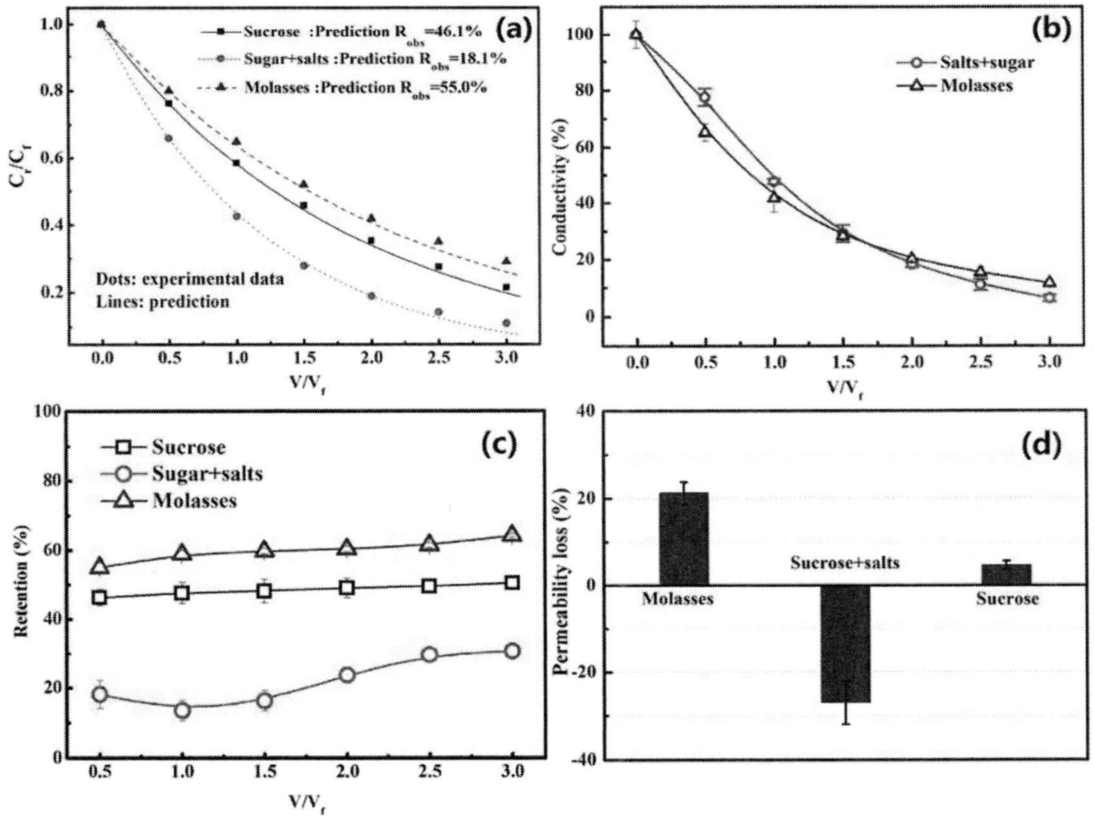


**Fig. 8** Variations of **a** sucrose concentration and **b** conductivity in the retentate during diafiltration of different feed solutions; **c** calculated sucrose retention variation and **d** permeability loss for the diafiltration of different feed solutions by NF5 membrane. Operating parameters: 60 °C, 17.26 $Lm^{-2}$ $h^{-1}$, 1200 rpm, $V/V_f = 3$

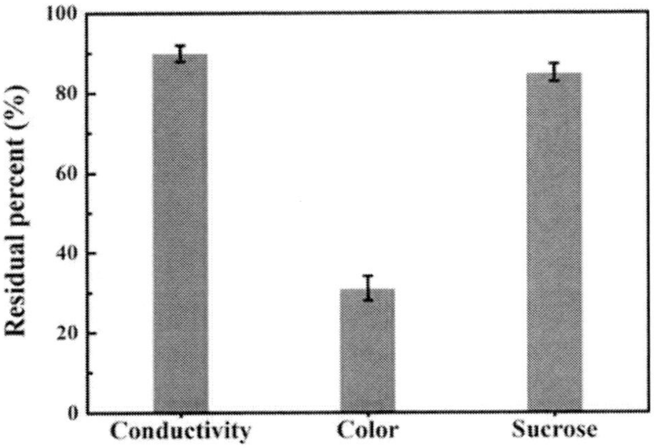


**Fig. 9** Residual components of molasses after adsorption by resin. Operating parameters: diluted molasses (water/molasses = 1:3) as feed, 60 °C, HPD722-resin, 120 g/100 mL (resin/molasses)

membrane surface kept clean for the NF5 membrane. According to the results in Fig. 4, the permeability loss indeed occurred for the NF5 membrane with molasses, implying that the pore narrowing or blocking was its main fouling mechanism.

Theoretically, the salt concentration was decreasing and the fouling was growing during the diafiltration process. In order to further clarify the long-term effect of salt and fouling on the sucrose retention, longer diafiltration process ($V/V_f = 3$) with the NF5 membrane was carried out and the sucrose retention was monitored. As shown in Fig. 8, the sucrose concentration in the retentate decreased much faster for the sugar-salt mixture than diluted molasses, whereas their salt concentration decline rate was similar, indicating that the "salting-out" effect and membrane fouling affected the sucrose retention rather than the salt retention. Moreover, the deviation between prediction lines and experimental results was observed for all the solutions because the sucrose retention was varying with diafiltration volume (Fig. 8a, c). For the sugar-salt mixture, the fluctuant sucrose retention could be explained as follows: at the beginning of diafiltration, the "salting-out" effect happened and exhibited hysteresis, and with decreasing salt concentration during diafiltration, such salt effect was alleviated thus resulting in a slow increase in the sucrose retention. This special phenomenon was not found for the diafiltration of molasses, implying that the pore swelling and sucrose dehydration disappeared when in the presence of pigments, and the increase of sucrose retention for molasses was mainly caused by gradual fouling evolution (Fig. 8d, where the permeability loss is much higher than that in Fig. 4b). This also confirmed that the pigments could mitigate or even eliminate the pore swelling and sucrose dehydration mainly by fouling formation. For single sucrose solution, the increase of sucrose retention was mainly resulted from the weaker concentration polarization (sucrose concentration was decreasing with diafiltration) because its irreversible fouling was less than 5%.

It can be speculated that the non-polar pigments (less hydrophilic) in molasses are easier to loss their hydration layer or adsorb in the membrane compared to the polar pigments (charged and more hydrophilic). We hypothesize that the non-polar pigments counteract the pore swelling and sucrose dehydration by competitive dehydration or fouling neutralizing (Fig. 5), thus increasing the sucrose retention. In order to minimize the sucrose retention by the NF5 membrane, this kind of pigments should be removed in order to reappear the pore swelling and sucrose dehydration. Accordingly, the diluted molasses was pretreated by a non-polar macroporous resin (HPD722-resin) and most of the non-polar pigments were removed while other components were almost remained (Fig. 9, polar pigments accounting for 30% of the total color were remained). It was worth mentioning that this resin could be completely regenerated by ethanol washing and ethanol was able to be reused by evaporation-condensation (avoid discharge of hypersaline wastewater). As seen in Fig. 10, the sucrose retention by the NF5 membrane decreased obviously and the membrane permeability increased again after the filtration with the pretreatment by non-polar resin. This confirmed that the non-polar pigments

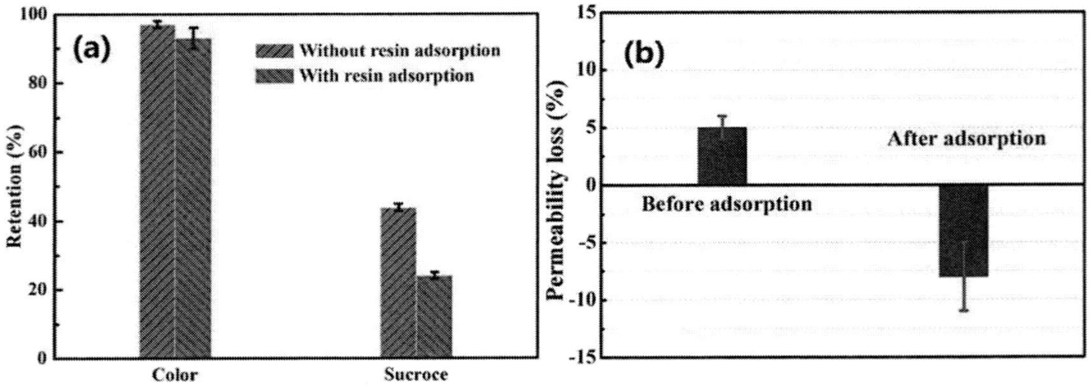


**Fig. 10** Filtration behavior of diluted molasses with and without nonpolar pigments by NF5 membrane: **a** color and sucrose retention; **b** permeability loss. Operating parameters: 60 °C, 17.26 $Lm^{-2}$ $h^{-1}$, 1200 rpm, $V/V_f = 1$

were responsible for the disappearing of pore swelling and sucrose dehydration, and the extraction of these pigments by the macroporous resin could not only obtain a highly valued product (caramel color), but also reduce membrane fouling and increase sucrose permeation.

## Effect of Operating Conditions

As discussed above, the removal of the non-polar pigments could improve sucrose permeation through NF membrane by reappearing the pore swelling and sucrose dehydration. It was also well known that the operating conditions (e.g., temperature, permeate flux, and shear rate on the membrane) affected solute retention and membrane fouling (Zhu et al. 2018; Zhang et al. 2015), which could be optimized to maximize sucrose permeation through NF5 in the following studies.

Higher temperature increases membrane pore size (Fig. 3b) and accelerates solute diffusion through the membrane (Manttari et al. 2002). As seen in Fig. 11a, with increase of temperature from 40 to 60 °C, the color retention kept above 95% while the sucrose retention decreased from 76 to 56%. This indicated that high temperature could improve the separation of pigments and sucrose. Moreover, the TMP during diafiltration was lower at higher temperature although the irreversible fouling was slightly increasing with temperature (Fig. 12a). The lower TMP was caused by the membrane pore enlargement, weaker concentration polarization (due to stronger back diffusion of solutes), and lower solute retention (smaller osmotic pressure difference across the membrane) at higher temperature, whereas the higher irreversible fouling could be explained by the increasing pore blockage due to the temperature-induced membrane pore enlargement and higher solute diffusion (Luo et al. 2016). Therefore, if the membrane fouling could be controlled at higher temperature, the sucrose permeation would be further promoted.

It was well known that permeate flux has great effect on the solute retention and membrane fouling due to convection-diffusion mechanism (Luo and Wan 2013). As shown in Fig. 11b and Fig. 12b, the retention for both pigment and sucrose increased with permeate flux, especially for sucrose, from 37 to 59%. Moreover, the TMP and irreversible fouling also increased as expected at higher permeate flux, indicating that the higher sucrose retention was not only caused by the "dilution effect" (more convection transport of solvent than solutes) (Luo and Wan 2013) but also the pore narrowing effect (Goulas et al. 2002). Although lower permeate flux could benefit sucrose recovery and fouling control, more membrane area was required to treat the same volume of molasses, thus increasing the equipment disposable investment.

Increasing the shear rate at membrane surface can enhance the back-transport of solutes and alleviate concentration polarization, thus improving solute retention and decreasing membrane fouling (Luo et al. 2018). However, as seen in Fig. 11c, when the stirring speed in the dead-end filtration cell ascended from 200 to 1200 rpm, the retention of pigment did not change significantly, and the sucrose retention increased a little when the stirring speed raised to 1700 rpm, implying that concentration polarization at high temperature (60 °C) and

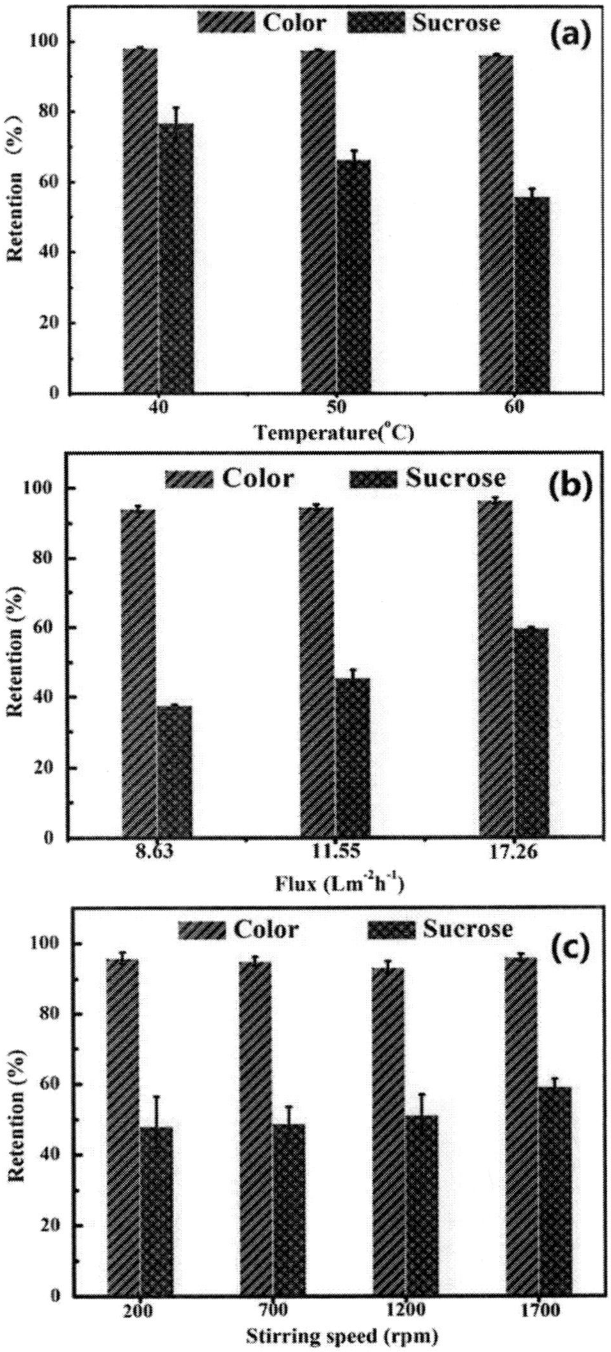


**Fig. 11** Effect of **a** temperature, **b** permeate flux, **c** stirring speed on the retention behavior of NF5 membrane. Operating parameters: diluted molasses as feed solution, 17.26 $Lm^{-2} h^{-1}$, 1200 rpm, $V/V_f = 1$

low permeate flux (17.26 $Lm^{-2} h^{-1}$) was not severe (Djurić et al. 2004). Thus, the effect of stirring speed on the TMP during the diafiltration was not important and their TMP difference was below 1 bar probably due to the permeability dissimilarity of membrane disks (Fig. 12c), whereas the irreversible fouling was slightly decreasing from 15 to 8% with increase of stirring speed because higher shear rate at the membrane surface prevented foulant deposition. But, the benefits resulted from fouling control may not offset the energy cost for providing such high shear in this case. .

As discussed above, temperature and flux were the two critical parameters influencing the membrane performance, and thus, the interaction of the two factors on the sucrose filtration behavior was investigated using single sucrose model solution. It was clearly shown in Fig. 13a that the sucrose retention increased with permeate flux due to the "dilution

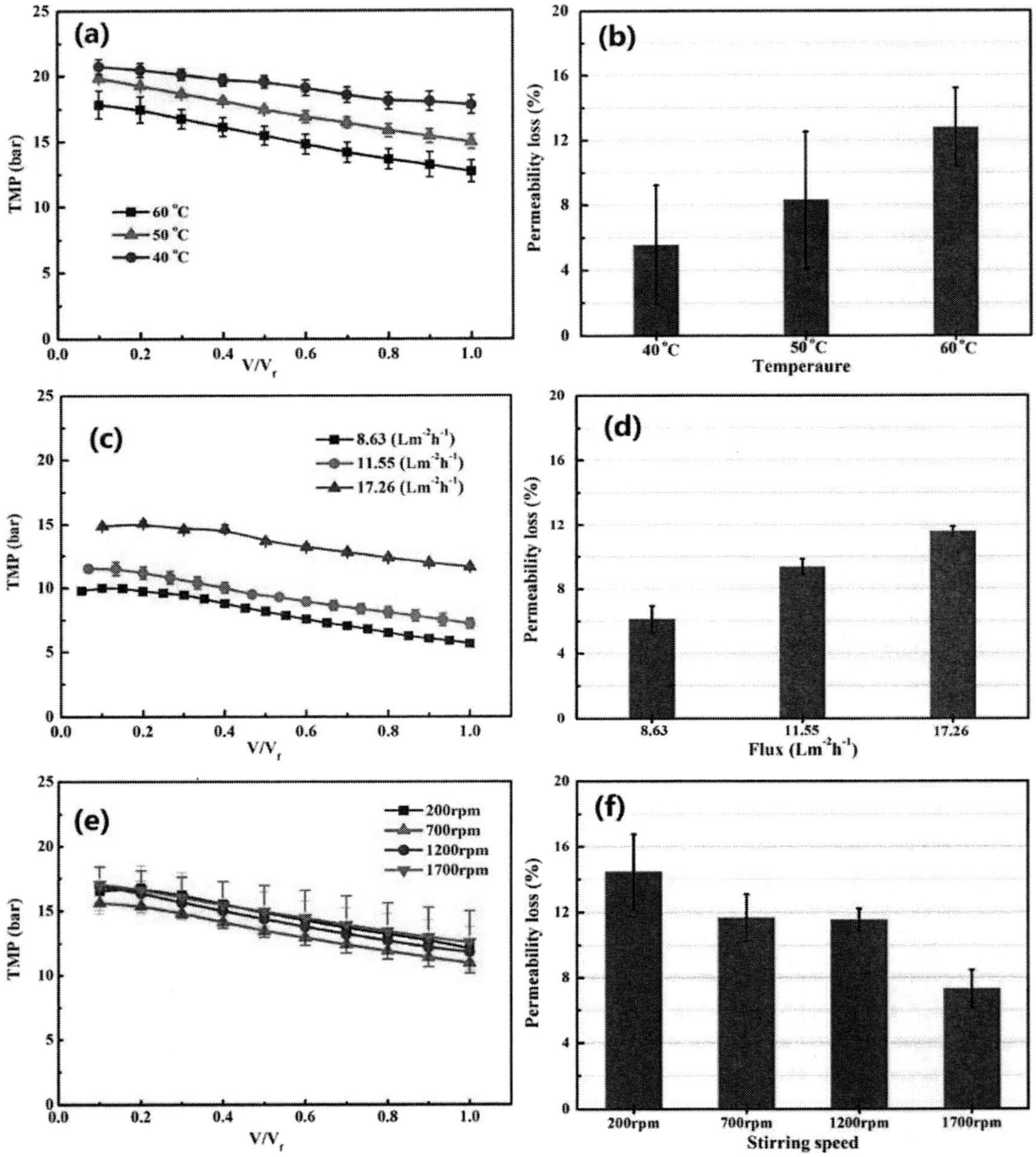


**Fig. 12** Effect of temperature on **a** TMP variation and **b** irreversible fouling, effect of permeate flux on **c** TMP variation and **d** irreversible fouling, effect of stirring speed on **e** TMP variation and **f** irreversible fouling of NF5 membrane during diafiltration of diluted molasses. Operating parameters: 60 °C, 17.26 $Lm^{-2} h^{-1}$, 1200 rpm, $V/V_f = 1$

effect" but the slope of their linear fitting was smaller at lower temperature. This could be explained by a fact that at lower temperature, concentration polarization became more serious due to smaller diffusion coefficient of solute, and the resulting higher sucrose concentration at the membrane surface would bring more sucrose passing through the membrane, which could partly offset the "dilution effect" induced by higher permeate flux and retard the resulting retention increase. This could be verified by the calculated results in Fig. 13b based on film theory (Eq. 7), showing that the ratio of sucrose concentration at membrane surface to its feed concentration was greater at lower temperature and higher permeate flux. Therefore, higher temperature could weaken the increase in concentration polarization induced by the higher permeate flux, which was normally considered a positive effect in membrane filtration. Here, it was not helpful for sucrose permeation in this regard although higher temperature could boost sucrose permeation by enlarging membrane pore and increasing solute diffusion transport across the membrane. Moreover, higher temperature decreased the sucrose concentration difference (ΔC) across the membrane (Fig. 13c), thus reducing the osmotic pressure difference (ΔΠ) (Fig. 13d), and this would save lots of energy by diminishing the operating pressure. However, if comparing the results in Fig. 13c, d, it was found that the temperature-induced difference in osmotic pressure became smaller than that in solute concentration, because higher temperature also increased the pure osmotic pressure of solute as indicated in Eq. 9.

## Conclusions

This work unraveled the mechanism of high sucrose retention during decoloration of molasses with UF and NF membranes by systematically investigating the effect membrane properties, molasses component, and operating parameters on the sucrose retention behaviors. For PES membranes, dense fouling layer was easy to form and greatly improved the color removal/sucrose retention, and the UP005 membrane seemed suitable for this purpose, whereas for PA membranes, thanks to their high surface hydrophilicity, a self-made NF membrane with a MWCO of 300–500 Da outperformed other membranes tested. However, RC membranes could not be employed for the molasses decoloration due to their low color

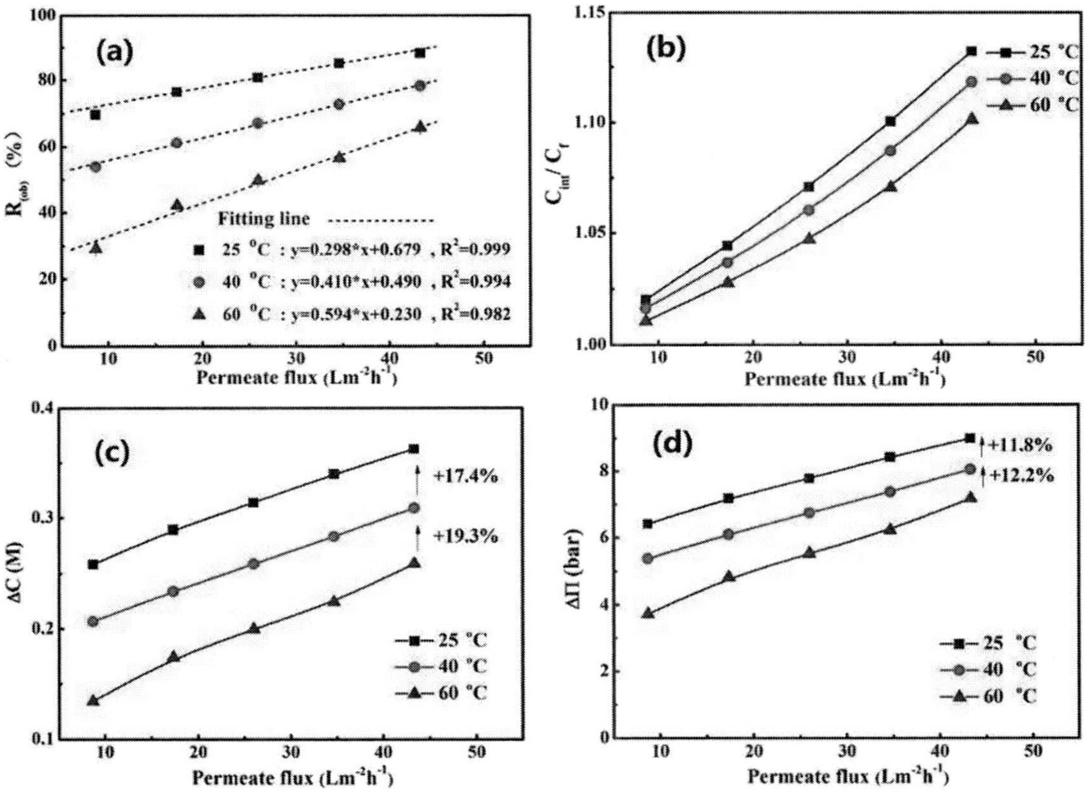


**Fig. 13** Effect on permeate flux and temperature on the filtration behavior of NF5 membrane using model sucrose solution as feed: **a** observed sucrose retention, **b** ratio of sucrose concentration on membrane surface to its bulk concentration, **c** sucrose concentration difference across the membrane, **d** osmotic pressure difference across the membrane. Operating parameters: sucrose concentration 150 g $L^{-1}$, 1200 rpm, $V/V_f = 0.5$

removal and the membrane swelling induced by high salt concentration and temperature.

Reducing sugar in molasses produced little effect on the sucrose retention, while inorganic salts resulted in pore swelling and sucrose dehydration due to "salting-out" effect, thus improving the sucrose permeation through the membrane. However, the presence of pigments in molasses would weaken or eliminate such positive effect mainly by adsorption fouling formation, thus recover or increase the sucrose retention. Based on the underlying mechanisms, the removal of non-polar pigments by macroporous resin could improve sucrose permeation through NF membrane by reappearing the pore swelling and sucrose dehydration. High temperature could accelerate the sucrose permeation across the membrane, but it also attenuated concentration polarization layer, thus intensifying the sucrose retention increase induced by high permeate flux. Therefore, efficient decoloration of molasses with high sucrose permeation could be achieved by selecting suitable membrane, manipulating fouling formation, reappearing salt effect on membrane and sucrose, as well as operating at higher temperature and lower permeate flux. These obtained knowledge also can be extrapolated to the decoloration of other bioproduct raw materials by membrane filtration.

**Funding Information** The financial supports are supplied by Guangxi Science and Technology Major Project (AA17204034) and Key Research Program of Chinese Academy of Sciences (KFZDSW-211-3). This work is also supported by "100 Talents Program" of Chinese Academy of Sciences.

## References

Ambrosi, A., Medeiros Cardozo, N. S., & Tessaro, I. C. (2014). Membrane separation processes for the beer industry: A review and state of the art. *Food and Bioprocess Technology., 7*(4), 921–936.

Bargeman, G., Westerink, J. B., Mignez, O. G., & Wessling, M. (2014). The effect of NaCl and glucose concentration on retentions for nanofiltration membranes processing concentrated solutions. *Separation and Purification Technology., 134*, 46–57.

Bernal, M., Ruiz, M. O., Geanta, R. M., Benito, J. M., & Escudero, I. (2016). Colour removal from beet molasses by ultrafiltration with activated charcoal. *Chemical Engineering Journal., 283*, 313–322.

Blanc, C.-L., Lemaire, J., Duval, F., Theoleyre, M.-A., & Pareau, D. (2017). Purification of pentoses from hemicellulosic hydrolysates without neutralization for sulfuric acid recovery. *Separation and Purification Technology., 174*, 513–519.

Bowen, W. R., & Jenner, F. (1995). Theoretical descriptions of membrane filtration of colloids and fine particles: An assessment and review. *Advances in Colloid & Interface Science., 56*(13), 141–200.

Bowen, W. R., Mohammad, A. W., & Hilal, N. (1997). Characterisation of nanofiltration membranes for predictive purposes — Use of salts, uncharged solutes and atomic force microscopy. *Journal Of Membrane Science., 126*(1), 91–105.

Boy, V., Roux-de Balmanna, H., & Galier, S. (2012). Relationship between volumetric properties and mass transfer through NF membrane for saccharide/electrolyte systems. *Journal of Membrane Science., 390*, 254–262.

Colla, E., Pereira, A. B., Hernalsteens, S., Maugeri, F., & Rodrigues, M. I. (2010). Optimization of Trehalose production by Rhodotorula dairenensis following a sequential strategy of experimental design. *Food and Bioprocess Technology., 3*(2), 265–275.

Djurić, M., Gyura, J., & Zavargo, Z. (2004). The analysis of process variables influencing some characteristics of permeate from ultra- and nanofiltration in sugar beet processing. *Desalination, 169*(2), 167–183.

Escoda, A., Fievet, P., Lakard, S., Szymczyk, A., & Deon, S. (2010). Influence of salts on the rejection of polyethyleneglycol by an NF organic membrane: Pore swelling and salting-out effects. *Journal of Membrane Science., 347*(1–2), 174–182.

Goulas, A. K., Kapasakalidis, P. G., Sinclair, H. R., Rastall, R. A., & Grandison, A. S. (2002). Purification of oligosaccharides by nanofiltration. *Journal of Membrane Science., 209*(1), 321–335.

Gyura, J., Seres, Z., & Eszterle, M. (2005). Influence of operating parameters on separation of green syrup colored matter from sugar beet by ultra- and nanofiltration. *Journal of Food Engineering., 66*(1), 89–96.

Hai, F. I., Yamamoto, K., Nakajima, F., & Fukushi, K. (2011). Bioaugmented membrane bioreactor (MBR) with a GAC-packed zone for high rate textile wastewater treatment. *Water Research, 45*(6), 2199–2206.

Hamachi, M., Gupta, B. B., & Ben Aim, R. (2003). Ultrafiltration: A means for decolorization of cane sugar solution. *Separation and Purification Technology., 30*(3), 229–239.

Liu, M., Zhu, H., Dong, B., Zheng, Y., Yu, S., & Gao, C. (2013). Submerged nanofiltration of biologically treated molasses fermentation wastewater for the removal of melanoidins. *Chemical Engineering Journal., 223*, 388–394.

Luo, J., Ding, L., Su, Y., Wei, S., & Wan, Y. (2010). Concentration polarization in concentrated saline solution during desalination of iron dextran by nanofiltration. *Journal of Membrane Science., 363*(1–2), 170–179.

Luo, J., Guo, S., Wu, Y., & Wan, Y. (2018). Separation of sucrose and reducing sugar in cane molasses by Nanofiltration. *Food and Bioprocess Technology., 11*(5), 913–925.

Luo, J., Hang, X., Zhai, W., Qi, B., Song, W., Chen, X., & Wan, Y. (2016). Refining sugarcane juice by an integrated membrane process: Filtration behavior of polymeric membrane at high temperature. *Journal of Membrane Science., 509*, 105–115.

Luo, J., & Wan, Y. (2013). Effects of pH and salt on nanofiltration-a critical review. *Journal of Membrane Science., 438*, 18–28.

Mai, T. K., Rodtong, S., Baimark, Y., Rarey, J., & Boontawan, A. (2018). Membrane-based purification of optically pure D-lactic acid from fermentation broth to poly(D-lactide) polymer. *Journal of Membrane Science., 551*, 180–190.

Manttari, M., Pihlajamaki, A., Kaipainen, E., & Nystrom, M. (2002). Effect of temperature and membrane pre-treatment by pressure on the filtration properties of nanofiltration membranes. *Desalination, 145*(1–3), 81–86.

Miller, G. L. (1959). Use of Dinitrosalicylic acid reagent for determination of reducing sugar. *Analytical Chemistry., 31*(3), 426–428.

Miyagi, A., Suzuki, T., Nabetani, H., & Nakajima, M. (2013). Color control of Japanese soy sauce (shoyu) using membrane technology. *Food and Bioproducts Processing., 91*(C4), 507–514.

Moreno-Vilet, L., Bonnin-Paris, J., Bostyn, S., Ruiz-Cabrera, M. A., & Moscosa-Santillan, M. (2014). Assessment of sugars separation from a model carbohydrates solution by nanofiltration using a design of experiments (DoE) methodology. *Separation and Purification Technology., 131*, 84–93.

Nguyen, D. M. T., Bartley, J. P., & Doherty, W. O. S. (2017). Combined Fenton oxidation and chemical coagulation for the treatment of Melanoidin/phenolic acid mixtures and sugar juice. *Industrial & Engineering Chemistry Research., 56*(6), 1385–1393.

Opong, W. S., & Zydney, A. L. (1991). Diffusive and convective protein transport through asymmetric membranes. *Aiche Journal., 37*(10), 1497–1510.

Qi, B., Wu, Y., Guo, S., Luo, J., & Wan, Y. (2017). Refinement of cane molasses with membrane technology for Clarifcation and color removal. *Journal of Membrane Science and Research., 3*(4), 303–307.

Russin, T. A., Boye, J. I., Arcand, Y., & Rajamohamed, S. H. (2011). Alternative techniques for defatting soy: A practical review. *Food and Bioprocess Technology., 4*(2), 200–223.

Satyawali, Y., & Balakrishnan, M. (2008). Wastewater treatment in molasses-based alcohol distilleries for COD and color removal: A review. *Journal of Environmental Management., 86*(3), 481–497.

Schmidt, J. M., Greve-Poulsen, M., Damgaard, H., Hammershøj, M., & Larsen, L. B. (2016). Effect of membrane material on the separation of proteins and polyphenol oxidase in ultrafiltration of potato fruit juice. *Food and Bioprocess Technology., 9*(5), 822–829.

Sguarezi, C., Longo, C., Ceni, G., Boni, G., Silva, M. F., Di Luccio, M., Mazutti, M. A., Maugeri, F., Rodrigues, M. I., & Treichel, H. (2009). Inulinase production by agro-industrial residues: Optimization of pretreatment of substrates and production medium. *Food and Bioprocess Technology., 2*(4), 409–414.

Sharma, M., Patel, S. N., Lata, K., Singh, U., Krishania, M., Sangwan, R. S., & Singh, S. P. (2016). A novel approach of integrated bioprocessing of cane molasses for production of prebiotic and functional bioproducts. *Bioresource Technology., 219*, 311–318.

Subramaniam, M. N., Goh, P. S., Lau, W. J., Tan, Y. H., Ng, B. C., & Ismail, A. F. (2017). Hydrophilic hollow fiber PVDF ultrafiltration membrane incorporated with titanate nanotubes for decolourization of aerobically-treated palm oil mill effluent. *Chemical Engineering Journal., 316*, 101–110.

Tanninen, J., Manttari, M., & Nystrom, M. (2006). Effect of salt mixture concentration on fractionation with NF membranes. *Journal of Membrane Science., 283*(1–2), 57–64.

Wang, X. L., Zhang, C. H., & Ouyang, P. (2002). The possibility of separating saccharides from a NaCl solution by using nanofiltration in diafiltration mode. *Journal of Membrane Science., 204*(1–2), 271–281.

Xu, J.-J., Li, Q., Cao, J., Warner, E., An, M., Tan, Z., Wang, S.-L., Peng, L.-Q., & Liu, X.-G. (2016). Extraction and enrichment of natural pigments from solid samples using ionic liquids and chitosan nanoparticles. *Journal of Chromatography A., 1463*, 32–41.

Zhang, W., Luo, J., Ding, L., & Jaffrin, M. Y. (2015). A review on flux decline control strategies in pressure-driven membrane processes. *Industrial & Engineering Chemistry Research., 54*(11), 2843–2861.

Zhao, L., Zhao, H., Phuongbinh, N., Li, A., Jiang, L., Xia, Q., Rong, Y., Qiu, Y., & Zhou, J. (2013). Separation performance of multi-components solution by membrane technology in continual diafiltration mode. *Desalination, 322*, 113–120.

Zhou, Y., Lai, Y. S., Eustance, E., Straka, L., Zhou, C., Xia, S., & Rittmann, B. E. (2017). How myristyltrimethylammonium bromide enhances biomass harvesting and pigments extraction from Synechocystis sp PCC 6803. *Water Research, 126*, 189–196.

Zhu, Z., Luo, X., Yin, F., Li, S., & He, J. (2018). Clarification of Jerusalem artichoke extract using ultra-filtration: Effect of membrane pore size and operation conditions. *Food and Bioprocess Technology., 11*(4), 864–873.

# 522027 Energy-Water Nexus Study for a Mushroom Farming Initiative in Nigeria

**Abstract:**

A water treatment and recycle system was designed for a mushroom farming initiative in Kaduna, Nigeria. The energy requirements of the proposed system were studied to determine the optimal system to minimize energy use for water treatment. Other considerations were ease of use and capital and operating costs.

The concept behind the cooperative is for 200 women farmers pool their resources to pay for water and energy costs. Electricity will be provided through solar panels. We were consulted to design a water treatment and recycle system for a demonstration of the concept. The concept will likely be replicated in remote areas without an abundance of water, so water recycle is essential. Reliable electricity will also likely be an issue so solar panels will be used.

The first step was to collect samples of water from the potential water source: the Kaduna River. Preliminary tests showed that the water was relatively clean but contained bacteria and some hardness. A treatment and recycle system was designed appropriate for the impurities in the water. Due to limited energy provided by the solar panels, energy requirements were studied for all appropriate treatments to minimize energy use of the final system. The final design and final energy use of the water system will be presented.

# 523808 Sustainable Optimal Strategic Planning for Shale Water Management


**Abstract:**

The development of new additives in the shale gas industry, which tolerate the use of high Total Dissolved Solids (TDS) base fluid, has allowed reusing the wastewater in the drilling of subsequent wells (U.S. Environmental Protection Agency, 2016). This practice is currently the most popular and cost-effective option for shale gas water management. Although it minimizes freshwater consumption, producers should take into consideration possible long-term issues and challenges. For instance, the TDS concentration will increase significantly, around 2-3 times, which can represent a cost barrier to reuse the water for fracturing operations. Besides, as the number of drilled wells decrease this practice becomes less attractive. Specifically, the volume of fracturing fluid required to fracture new wells may be less than the volume of water generated by producing wells in the area.

Currently, several works have been reported on the optimization of shale gas water management (Gao and You, 2015; Guerra et al., 2016; Yang et al., 2015). However, most of the works published in the literature consider that the water blending ratio is restricted or the return to pad operations are not allowed. Drouven and Grossmann (2017) assume that the water-blending ratio is unrestricted, they over-estimate the friction reducers expenses. The MILP model that they proposed does not account the salt concentration of impaired water. Moreover, they only distinguish between two types of water: impaired water and freshwater. Hence, the model cannot handle any water management option when drilling operation decrease.

In this work, we propose a mixed-integer non-linear programming (MINLP) model considering the TDS concentration of flowback and impaired water as a function of time and for different water treatment solutions. We estimate the friction reducers expenses as a function of TDS concentration to determine if the level of TDS in impaired water is an impediment to reusing it in fracturing operations. Moreover, the model distinguishes between four types of water: impaired water, flowback water, desalinated water and freshwater. The objective is to maximize the "sustainability profit" in order to obtain a compromise solution between economic, environmental and social aspects. Only a single objective function is necessary since all the indicators are expressed in monetary terms (Zore et al., 2017).

The water management system comprises wellpads, shale gas wells in each wellpad, centralized water treatment technologies (CWT), freshwater sources, fracturing crews, and disposals wells.

After hydraulic fracturing, a portion of water called flowback water returns to the wellhead. The flowback water is stored in fracturing tanks (FT) onsite before basic treatment (pre-treatment) in mobile units or transport to CWT facility or Class II disposal. Pre-treatment includes technologies to remove suspended solids, oil and grease, and bacteria, certain ions that can cause the scale to form on equipment and interfere with fracturing chemical additives. After pre-treatment, the water can be used as a fracturing fluid in the same wellpad or in the neighboring wellpad, or can be desalinated in the onsite TDS removal technologies. The flowback water reused for fracturing operations is called impaired water and is stored in impaired water tanks (IWT). Several desalination technologies can be selected such as multi-stage membrane distillation (MMD), multi-effect evaporation with mechanical vapor recompression (MEE-MVR) or forward-reverse osmosis (FO-RO) hybrid. We consider that the outflow brine salinity in the onsite treatment is close to salt saturation conditions to achieve ZLD operation. Cost and salinity levels restrict the type of desalination unit that can be used for TDS removal. The onsite desalinated water can also be used as a fracturing fluid in the same wellpad, transported to the next wellpad or discharge for other usages. The flowback water can also be transported and treated in CWT. Desalinated water from CWT can select the same routes as the desalinated water in onsite technologies. Freshwater is obtained from an uninterruptible fresh water source. Desalinated water and freshwater are stored in freshwater tanks (FWT) and/or water impoundment. Transportation of freshwater, impaired water, flowback water and desalinated water can only be through trucks. Storage tanks and mobile treatment are assumed to be leased.

The problem is to determine: (1) wellpad fracturing start date (fracturing schedule), (2) number of tanks leased at each time period, (3) number of trucks needed in each time period, (4) flowback destination, reuse (impaired water) , treatment (onsite or offsite) or disposal, and (5) quality of water used to fracture each well.

The optimization problem is formulated as an MINLP model that includes: assignment constraints, material balance in storage tanks, mixer and splitters, logic constraints and an objective function. Salt material balances are modeled using total flows and salt composition as variables (bilinear terms). An advantage of using this representation is that the variables involved in the bilinear terms are well bounded, allowing us to define tight under and over estimators.

Them objective function includes the economic profit, the eco-profit and the social-profit.

- *Economic profit* includes revenues from natural gas minus the sum of the following expenses: wastewater disposal cost, storage tank cost, freshwater cost, friction reducer cost, wastewater and freshwater transport cost and onsite and offsite treatment cost.
- *Eco-profit* distinguishes between eco-benefit (raw material and products that unburden the environment) and eco-cost (raw material al products that burden the environment). Both terms are calculated by using eco-cost coefficients (Delft University of Technology, 2017). In our problem, impaired water and desalinated water used to fracture a neighboring well exhibit unburdening effect on the environment. However, natural gas, freshwater withdrawal, disposal and transportation burden the environment.
- *Social profit* includes social security contributions paid for the employed people to fracture a well, plus the social transfer by hiring people, minus social cost. We only take into account the numbers of jobs on a fracturing crew and the time that they are working to fracture a specific well. Once the well is completed, the number of jobs generated by truck drivers or maintenance team are not considered.

The proposed model is applied to a case study in Marcellus Play with 3 wellpads and 20 wells, three years discretized at one week per time period, four interruptible sources of fresh water, three class II disposal wells, two CWT plants and one fracturing crew. The MINLP problem consists of 2808 binary variables, 14977 continues variables and 12010 constraints.

The solution of the model shows that the capabilities to reuse impaired water to fracture other wells is the best economic and environmental practice for shale gas water management. The model, which can be effectively solved with the proposed decomposition technique, reveals that the level of TDS in impaired water is not an obstacle to reusing it for fracturing purposes. Also, it has been shown that onsite desalination treatment can be cost-effective for operators once no more wells to fracture are available.

**References:**

Delft University of Technology, 2017. The Model of the Eco-costs/Value Ratio (EVR). URL http://www.ecocostsvalue.com (accessed 12.1.17).

Drouven, M.G., Grossmann, I.E., 2017. Optimization models for impaired water management in active shale gas development areas. J. Pet. Sci. Eng. 156, 983–995.
Gao, J., You, F., 2015. Shale Gas Supply Chain Design and Operations toward Better Economic and Life Cycle Environmental Performance: MINLP Model and Global Optimization Algorithm. ACS Sustain. Chem. Eng. 3, 1282–1291.
Guerra, O.J., Calderón, A.J., Papageorgiou, L.G., Siirola, J.J., Reklaitis, G. V., 2016. An optimization framework for the integration of water management and shale gas supply chain design. Comput. Chem. Eng. 92, 230–255.
U.S. EPA, 2016. Technical Development Document For Effluent Limitations Guidelines and Standars for the Oil and Gas Extraction Point Source Category. Washington, DC.
Yang, L., Grossmann, I.E., Mauter, M.S., Dilmore, R.M., 2015. Investment optimization model for freshwater acquisition and wastewater handling in shale gas production. AIChE J. 61, 1770–1782.
Zore, Ž., Čuček, L., Kravanja, Z., 2017. Syntheses of sustainable supply networks with a new composite criterion – Sustainability profit. Comput. Chem. Eng. 102, 139–155.

# 534130 A Multi-Objective Energy-Water Nexus Planning Model: A Case Study of the Power Systems in Texas Edwards Aquifer

**Abstract:**

Increasing power demands and water scarcity as well as environmental concerns give rise for the need of a planning strategy for power generating systems that takes into consideration the nexus between energy and water [1]. Power systems planning problems are complex and comprise of decisions related to: (i) the ability to add additional power plants with different types of cooling and power generating technologies, (ii) the ability to convert the cooling and or power generating technologies of existing power plants, (iii) uncertainties, and (iv) environmental and social impacts [2]. Given that the type of power generating and cooling technology of a power plant directly affects its water usage, the optimization of a power system is a multi-objective problem [3]. Therefore, a systematic methodology based on the Energy-Water Nexus (EW-N) is necessary for modeling and solving the optimization problem on power systems planning [4].

In this work, we develop a novel superstructure for planning strategies of power generating systems based on EW-N connections. The EW-N problem is formulated as a two-stage stochastic multi-objective mixed integer linear program [5]; with "here-and-now" decision variables such as additional power plants with specific power generation and cooling technologies and "wait-and-see" decision variables comprising of and not limited to the capacity and water utilization of power plants. The model determines the optimal conversion and expansion policies for the current infrastructure, based upon uncertain future water availability and power demands. The model is implemented into a case study regarding EW-N in the Edward Aquifer, aiming to provide policy makers with a systematic tool to analyze various scenarios and technology options. Moreover, the model is also flexible in that it can also be integrated into the Edwards Aquifer Groundwater and River System Simulation Model (EDSIMR) [6, 7], which is a regional model for the comprehensive decision-making and resources allocation regarding Food-Energy-Water Nexus in the Edwards Aquifer.

References:

[1] Diangelakis, N. A., & Pistikopoulos, E. N. (2017). A multi-scale energy systems engineering approach to residential combined heat and power systems. Computers & Chemical Engineering, 102, 128-138.

[2] Koltsaklis, N. E., Dagoumas, A. S., Kopanos, G. M., Pistikopoulos, E. N., & Georgiadis, M. C. (2014). A spatial multi-period long-term energy planning model: a case study of the Greek power system. Applied Energy, 115, 456-482.

[3] Mohtar, R. H., & Daher, B. (2016). Water-energy-food nexus framework for facilitating multi-stakeholder dialogue. *Water International*, *41*(5), 655-661.

[4] Garcia, D. J., & You, F. (2016). The water-energy-food nexus and process systems engineering: a new focus. *Computers & Chemical Engineering*, *91*, 49-67.

[5] Birge, J. R., & Louveaux, F. (2011). Introduction to stochastic programming. Springer Science & Business Media.

[6] Gillig, D., McCarl, B. A., & Boadu, F. (2001). An economic, hydrologic, and environmental assessment of water management alternative plans for the south central Texas region. Journal of Agricultural and Applied Economics, 33(1), 59-78.

[7] McCarl, B. A., Dillon, C. R., Keplinger, K. O., & Williams, R. L. (1999). Limiting pumping from the Edwards Aquifer: an economic investigation of proposals, water markets, and spring flow guarantees. Water Resources Research, 35(4), 1257-1268.

# 534167 Optimal Use of Thermal Membrane Distillation (TMD) for Treatment of Flowback Water

**Abstract:**

**Optimal use of thermal membrane distillation (TMD) for treatment of flowback water**

Kaiyu Cao[1], Priscille I. Etoughe[2], Rajib Mukherjee[2,3], Debalina Sengupta[3], Joseph S. Kwon[1,2,3*], Mahmud M. El-Halwagi[1,2,3]

1. *Artie McFerring Department of Chemical Engineering, Texas A&M Univrsity, College Station, TX*
2. *Texas A&M Energy Institute, Texas A&M University, College Station, TX*
3. *Gas and Fuels Research Center, Texas A&M Engineering Experiment Station, College Station, TX*

The flow rate of flowback water from hydraulic fracturing varies with time. Most of the water are produced within the first two weeks after hydraulic fracturing is completed. The concentration of total dissolved solids (TDS) in the flowback water also changes with time. For efficient recovery and reuse of flowback water, treatment with thermal membrane distillation (TMD) for the removal of TDS has been found to be the method of choice [1]. Even though flowback water is a continuous process, in this work, for modeling purpose a multi-period optimization problem has been formulated to account for the time-dependent variation of the flow rate and TDS concentration of flowback water, accounting for the time-; hereeach day is treated as a period. Then, we have integrated a TMD unit, considering the largest capacity of flowrate from a well in any given period. An upper limit on the concentration of TDS in the discharged (reject) water is considered to avoid precipitation. Utilizing the field data of flowback water from a given well for each period, the optimization formulation has been solved to determine the maximum period of time for optimal use of TMD, maximizing annualized profit from recovered water per period. Our result shows that it is not profitable to use water recovery beyond first few days, depending on the formation. The performance of the optimization formulation is measured in terms of percentage of recovery and is compared with cost associated with them.

**References:**

[1] Elsayed, N. A., Barrufet, M. A., Eljack, F. T., & El-Halwagi, M. M. (2015). Optimal design of thermal membrane distillation systems for the treatment of shale gas flowback water. *Int J Membr Sci Technol*, *2*, 1-9.

# 536214 The Energy-Water Nexus of Thermoelectric Power Generation and Its Impacts in the Muskingum River Watershed in Ohio

**Abstract:**

Thermoelectric power plants generate electricity and flue gas scrubbers mitigate emissions from the plants. However, at the same time the power plants withdraw large quantities of water from the watershed, which corresponds to 45% of the total 2010 US water use [1], and emissions from the scrubbers cause deposition of nutrients in the watershed. Therefore, increasing electricity generation could increase water stress and deteriorate water quality. Also, in the watershed where power plants are located, other activities, such as farming, interact with the power plants since these activities also require water as well as energy from the power plants and release nutrients to the watershed. To prevent shifting of environmental impacts across multiple flows [2], the Energy-Water nexus between different activities in the watershed needs to be understood in assessing the impacts of power plants.

Muskingum River Watershed (MRW) in Ohio ranks fourth as the most polluted watershed in the US [3], and two coal-fired power plants and three natural gas-fired power plants are located in the MRW. In this work, a holistic analysis approach is employed to evaluate the water-energy nexus in the MRW. Various technological and ecological systems are considered. Technological systems include mining, thermoelectric power generation, and $CO_2$ conversion activities. Ecological systems include land use and farming practice. Alternative scenarios for each component are analyzed to address the nexus. An agro-ecological model, which is developed by the Soil and Water Assessment Tool (SWAT), is used to assess water quality and quantity [4], while engineering models are used for the power plants and scrubbing processes.

Fossil fuel power plants not only require a huge amount of water, but also emit 28% of the 2016 US greenhouse gas emissions [5], 67% of the 2014 US $SO_2$ emissions, and 12% of the 2014 US $NO_x$ emissions [6]. Therefore, ecosystem services, such as water quality regulation, air quality regulation, and climate change regulation, play an important role in mitigating nutrient runoffs and air emissions. To account for the role of ecosystems, Techno-Ecological Synergy framework is applied in analyzing the impacts of activities in the MRW [7,8]. In this analysis, the trade-offs between water quality, water quantity, net electricity generation, carbon footprint, air quality, and net present value are identified. Various scenarios are considered to suggest better solutions that are 'win-win' in terms of multiple objectives.

Considering the Energy-Water nexus in the analysis will avoid shifting of the environmental impacts across multiple flows because interactions between flows are identified and all flows are included as objectives [2]. This work could be applied to any watershed and could be extended to address the Food-Energy-Water nexus by including food productions from farming activities as another objective. Addressing the nexus could help businesses and policy makers in making decisions to improve the sustainability of the watershed.

References

[1] Maupin, M.A. et al. Estimated use of water in the United States in 2010. U.S. Geological Survey, Reston, Virginia (2014).

[2] Bakshi, B. R., Gutowski, T. G. & Sekulic, D. P. Claiming Sustainability: Requirements and Challenges. ACS Sustain. Chem. Eng. 6, 3632–3639 (2018).

[3] Inglis, J. & Dutzik, T. Wasting our waterways: Toxic industrial pollution and restoring the promise of the clean water act. Environment America Research & Policy Center, Boston, MA (2006).

[4] Khanal, S., Lal, R., Kharel, G. & Fulton, J. Identification and classification of critical soil and water conservation areas in the Muskingum River basin in Ohio. J. Soil Water Conserv. 73, 213–226 (2018).

[5] Inventory of U.S. greenhouse gas emissions and sinks: 1990-2016. U.S. Environmental Protection Agency, Washington, D.C. (2018).

[6] 2014 National Emissions Inventory (NEI). US Environmental Protection Agency. https://www.epa.gov/air-emissions-inventories. Accessed on Apr 13 2018.

[7] Bakshi, B. R., Ziv, G. & Lepech, M. D. Techno-ecological synergy: A framework for sustainable engineering. Environ. Sci. Technol. 49, 1752–1760 (2015).

[8] Gopalakrishnan, V., Bakshi, B. R. & Ziv, G. Assessing the capacity of local ecosystems to meet industrial demand for ecosystem services. AIChE J. 62, 3319-3333 (2016).

# 537140 Systematic Analysis and Optimization of Water-Energy Nexus

**Abstract:**

Securing scarce energy and water resources from the environment, while at the same time satisfying on-growing water and energy demands for end-use consumption is crucial [1]. The intermediate processing facilities, energy sources (acting also as water sinks) and water sources (acting as energy sinks), receive energy and water natural resources respectively, process, exchange and deliver them as usable electricity, fuels and water for consumption. It is important to quantify the performance and the efficiency of this water-energy nexus, and understand how it is affected by the selection of the energy and water sourcesâ€™ technologies, operating capacities and their connectivity.

To this end, we present a scalable, graph-theoretic representation of a nexus for the design and optimization of complex water-energy systems [2]. We define a nexus as a directed bipartite graph with water and energy flows and show that for specified external grid demands, the optimal nexus configuration with minimum water and energy generation is the one without any redundant subgraphs. We then introduce the water-energy (WEN) diagram that can systematically identify and eliminate those redundant subsystems. This leads to (i) minimum generation/extraction of water and energy resources from the environment, or (ii) maximum yield of water and energy to meet external demands. Our approach is simple to implement and results in optimal nexus configurations that also consider operational constraints, connectivity restrictions and water quality specifications. In the second part, we develop mathematical programming-based algebraic method that provides the same results but for more complicated and large-scale nexus design. This also takes into account the intensity factors of the sources, restrictions on source-sink matching, and quality specifications. The model is further extended to include capital and operating costs of the processing facilities, as well as geographic distances. The approach is demonstrated using case studies on water-energy nexus systems focusing on power generation, seawater desalination, groundwater and surface water at regional and national scales.

**References:**

[1] D. J. Garcia and F. You, "The water-energy-food nexus and process systems engineering: a new focus," Computers & Chemical Engineering, vol. 91, pp. 49-67, 2016.

[2] S. Tsolas, M. N. Karim, M. M. F. Hasan, " Optimization of Water-Energy Nexus: A Network Representation-based Graphical Approach," Applied Energy, *Submitted.*

# 538380 Thermo-Economic Optimization Based Comparison of Membrane Distillation Vs Mechanical Vapor Recompression for Shale Gas Produced Water Treatment

**Abstract:**

With advancements in horizontal drilling and hydraulic fracturing methods, unconventional shale gas extraction is rapidly developing. However, shale gas production is accompanied with a host of environmental challenges. One such critical challenge is the sustainable management of large quantities of high salinity wastewater generated in the process of hydraulic fracturing. While recycling may be a short-term strategy, long term solutions are needed for economical and environmentally conscious management of high salinity wastewaters. Desalination is one possible technological solution to treat the produced water and reduce the brine volume for final disposal while simultaneously producing fresh water for other beneficial uses. Membrane distillation (MD) system is a vapor pressure driven emerging membrane technology capable of handling hypersaline wastewater. Since MD can be operated at low temperature (30-90$^0$C), it has the potential to be integrated with low-grade waste heat sources. Another advantage of MD is the low capital investment of the process. However, low energy efficiency of MD results in high operating costs when waste heat is not available. Mechanical vapor recompression (MVR) is another desalination technology capable of treating hypersaline produced water. Unlike MD, MVR system is a well-established work driven desalination process, in which the heat recovery and energy efficiency is relatively high by utilizing the vapor compression cycle. In contrast with MD, MVR system requires considerable amount of high quality energy sources. Considering the competitive desalination efficiency of these two technologies, there is a need for a systematic way to comparison their energetic and economic performance. To date, very limited work exists on techno-economic optimization and evaluation of MD and MVR technologies for produced water treatment.

We present a mixed-integer non-linear programming (MINLP) formulation of a direct contact membrane (DCMD) process design for produced water treatment while minimizing the total treatment cost. The proposed model simultaneously optimizes the operating conditions (i.e. the temperature, pressures, and flowrate of all streams) and DCMD plant configuration (number of membrane stages, number of modules in each stage, and membranes geometry). Also, the optimization model, takes into account the possibility to improve system performance by including brine recirculation and heat integration. The results of the proposed optimization model are compared with our previous work on MVR optimization for application in produced water treatment. The optimization model is tested for a hypothetical desalination plant with 0.5 million gallons per day (MGD) capacity, in which produced water with 10% salinity is concentrated to 30% salinity of the exiting brine. The results are also validated through ASPEN simulations. Sensitivity analysis is also conducted to analyze the effect of input parameters on the system performance and economics. Tradeoffs between MD and MVR technologies for treatment of shale gas produced water will be discussed.

# 529522 Application of Adsorbate Solid Solution Theory to Design Novel Adsorbents for Arsenic Removal Using Computer-Aided Molecular Design

**Abstract:**

**Application of Adsorbate Solid Solution Theory to Design Novel Adsorbents for Arsenic Removal Using CAMD**

Rajat K. Doshi[1,2], Arti Ayodhyaprasad Rajput[3], Rajib Mukherjee[1], Suresh Gupta[3] , Urmila M. Diwekar[1*],

1. Vishwamitra Research Institute, Crystal Lake, IL, USA
2. Henry B. Plant High School, 2415 S. Himes Ave., Tampa, USA
3. Department of Chemical Engineering, Birla Institute of Technology & Science, Pilani, India

*Corresponding: Urmila@vri-custom.org

Arsenic is a carcinogenic contaminant that pollutes the groundwater, a consequence of poor arsenic disposal. Various techniques are used to remove arsenic, such as oxidation, coagulation-flocculation, ion exchange, and adsorption, among which adsorption is the most efficient method. Current arsenic separating agents have a limited adsorption capacity. The overall objective of this work is to develop a computational tool for the design of novel adsorbents for arsenic remediation using clay materials including beidellite, zeolite, and sepiolite that are cheap and readily available [1]. In the first part of this research, we use the Group Contribution Method (GCM) to predict thermodynamic properties and calculate the UNIFAC interaction parameters between arsenic and other functional groups that are selected from clay materials. In the second part of this research, we utilize a computer-aided molecular design (CAMD) framework that develops new adsorbent candidates with enhanced adsorption capacities based on the group interaction parameters generated in the first part. The efficient ant colony optimization (EACO) algorithm maximizes the adsorption capacity with certain structural possibilities, thermodynamic property correlations, and process conditions. It was found that the newly designed adsorbent has an order of magnitude higher removal capacity than the adsorbents' reported in the literature [2]. The adsorption capacity estimated for the designed adsorbent was simulated in groundwater conditions; however, arsenic contamination is not solely confined to groundwater as it is also highly prevalent in wastewater flows from coal-burning power plants. An expansion of our work will be carried out by conducting experiments on the designed adsorbents. The goal is to test the performance of the designed adsorbent in coal power plant's wastewater flows and understand how the different ionic environment affects the adsorption capacity. The maximum amount of arsenic that the designed adsorbent can adsorbed per unit mass of adsorbent from wastewater found in coal-burning power plants will be presented. Moreover, the novel adsorbent has also been synthetically prototyped to determine the actual adsorption capacity of the novel adsorbent in varying ionic environments and arsenic concentrations. Our initial results from conducting a set of experimentation has demonstrated an enhanced adsorption capacity of the synthetic adsorbent that is higher than any naturally occurring adsorbent.

**References:**

1. Bektaş, N., Aydın, S., & Öncel, M. S. (2011). The adsorption of arsenic ions using beidellite, zeolite, and sepiolite clays: a study of kinetic, equilibrium and thermodynamics. Separation Science and Technology, 46(6), 1005-1016.
2. Doshi, R. K., Mukherjee, R., & Diwekar, U. M. (2018). Application of Adsorbate Solid Solution Theory To Design Novel Adsorbents for Arsenic Removal Using CAMD. ACS Sustainable Chemistry & Engineering, 6(2), 2603-2611.

# 538047 Analysis of a Circular Economy: From Food Waste to Foods

**Abstract:**

The US-EPA estimated that 38 million tons of food waste went to landfill in 2014 and in 2010 66.5 million tons of food didnâ€™t make it to consumers. The food waste streams total to around a billion pounds of food waste, where the majority go to landfill. Publicly owned waste water treatment plants in U.S. process approximately 14 billion kilograms of dry solids per year.[1] Turning waste streams into valuable products is the goal of the circular economy.[2]

Hydrothermal carbonization (HTC) process is proposed to be a solution to the large streams of food waste and biosolids from waste water treatment plants that creates a highly valuable, multiuse product. HTC is a thermochemical process that converts wet biomass materials to a coal-like product. Under sub-critical conditions, water in the HTC reactor stays in liquid form and acts as a reaction medium to promote the breakdown and cleavage of chemical bonds in the solid biomass. The absence of liquid-to-vapor phase change of the water makes the process significantly less energy intensive compared to a process that involves water vaporization or drying. The temperature range for HTC is from 180-240°C producing solid hydrochar as the main product.

The solid product of hydrothermal carbonization, hydrochar, has multiple applications ranging from a combustion fuel source, a catalyst for anaerobic digestion that increases the purity of methane production, a soil amendment, and the base for other carbon product such as activated carbon.

In this study we will explore the circular economy of utilizing organic wastes into foods. The steps involved in the circular economy system include: 1) collecting organic wastes (specifically food wastes and bio-solids), 2) converting the organic waste streams to hydrochar via HTC process, 3) growing food produces by indoor gardening, and 4) using hydrochar from step 2) as fuels to power the indoor gardens.

Utilizing organic waste streams in cities such as Chicago or Philadelphia, where there are a multitude of abandoned warehouses, provides an opportunity to produce locally grown produces that would typically come from across the country from California or Mexico. Not only will this circular economy reduce environmental impacts, but it will increase positive societal impacts by benefitting the local economy, providing educational opportunities of farming that would not typically be available in cities, and serving fresher produces to the local community.

A comparative life cycle assessment of our proposed process will be applied to compare conventional farming to our circular economy proposal. We hope to prove a reduction of environmental impacts, such as global warming potential, fossil fuel depletion, natural resource depletion. The life cycle assessment results will aid in a full system analysis which evaluate the proposed circular economy idea from societal, technical, environmental, economic, and political perspectives.

- https://www.epa.gov/sustainable-management-food/sustainable-management-food-basics
- https://www.sciencedirect.com/science/article/pii/S0301479717303808

*Contact: Dr. Justinus Satrio (justinus.satrio@villanova.edu)

# 538097 Re-Wiring the Domestic Food Trade for Reducing Irrigation Impacts in the United States

**Abstract:**

The food-energy-water(FEW) nexus concept recognizes the interconnected nature of the three systems and advocates managing them together to avoid sub-optimal outcomes and unintended consequences. The FEW nexus challenges differ for developed agro-economies such as the United States(U.S.) from agriculture centric developing nations. For U.S., irrigation provides a crucial link to study interconnections between food-water-energy resources. Additionally, food trade is an integral part of food system as it bridges the growing disconnect between production and consumption centers. Depending on the origin of food production, trade can potentially reduce environmental impact of food consumption. However, environmental impacts of agriculture are not directly considered when forming trade connections and may carry risk associated with suboptimal FEW system management (e.g. sourcing food from water-depleted regions). Structural optimization of food trade provides an opportunity to assess whether FEW resources can be managed together with respect to land-water-energy constraints. This work focuses on environmental impacts associated with FEW systems, namely- water required for irrigation, greenhouse gas(GHG) emissions from commodity transport, and pumping irrigation water for food production.

We leverage publicly available datasets from agriculture census and transportation surveys to create a network of regional food trade. The study combines data on state-wide food production, irrigation statistics, and life cycle inventory databases to convert food transfer network into embedded irrigation water, and GHG emission flows. Specifically, we focus on rice production and trade in the U.S. Rice is one of the staple food crops and produced entirely through irrigation, making it an ideal crop for the irrigation impact analysis. Majority of rice production is concentrated in four regions-Arkansas Grand Prairie, Mississippi Delta, Gulf Coast, and Sacramento Valley. The model optimizes rice trade network by keeping the demand constant and evaluate a series of constraints pertaining to reduction in energy and water usage to understand whether production can be shifted realistically to optimize all three systems. We limit the potential crop shifting to current production regions to account for soil and climatic parameters conducive to crop production. Additionally, we consider type of rice grown in each region as crop physiological requirements may differ based on the rice type. Preliminary results of the optimization framework reveals a rewired domestic rice trade network with a 27% reduction in virtual water use (15.1 billion $m^3$ reduced to 11.1 billion $m^3$). However, due to shift in production to more water-efficient states, a 10% increase in marginal land use is required. The land results are significant as they demonstrate the resulting sub-optimal solutions when interdependencies are overlooked. On the other hand, minimizing groundwater usage from water stressed areas also reduces overall GHG emissions demonstrating that these are not mutually exclusive goals in many states. The Pareto optimal solutions showing water-GHG emissions tradeoffs resulting from the model will be presented and their implications for regional FEW nexus sustainability will be discussed.

# 527168 Food, Energy, Fuels and Chemical Feedstocks from Rice Crops: Multi-Objective Optimisation of Multi-Product Value Chains for the Philippines

**Abstract:**

Rice is an important staple food in the Philippines such that its 20 M metric tons annual production places the country rank 8$^{th}$ among the rice producers in the world. However, there are many improvements in the efficiency and sustainability of the Philippines rice value chains that could be made by addressing various constraints related to technological, environmental, economic and socio-political factors [1]. Consequently, the Philippine government is striving to achieve Philippine rice competitiveness and to transform its value chains to be more profitable, resilient and sustainable; they recognise the importance of research and development as basis for such a laudable goal [2]. From a recent analysis of the Philippine rice value chain commissioned by the government, by-products like rice straw, rice hull and rice bran can be important sources of raw material for other industry sectors. Ninety-five percent of the rice bran is being used as animal feed component while seventy percent of the of the rice hull is being used as an alternative source of energy and fuel. Moreover, the silica-rich rice hull ash is being used as additive in the cement industry [3]. These could not only provide affordable renewable energy, fuels and materials but also much-needed rural development and employment opportunities. However, there are complex issues that need to be addressed: rice production is a water- and energy-intensive activity. The global water footprint from rice production is 1308 $Mm^3$/year with irrigated rice cultivation consuming the most water. Energy is required for the production of fertilisers, electricity for pumping water and heat for parboiling, drying and other milling activities [4]. Furthermore, the limited land for rice cultivation is being threatened by the rapid and unregulated conversion of agricultural land to other uses [1]. In this project, a comprehensive optimisation model, based on mixed-integer programming [5-6], is being developed that can support complex decision-making related to multi-product rice value chains and can capture the trade-offs between water, energy and land utilisation. The model is used to identify promising rice value chains for sustainable and efficient production of food, energy, fuels and chemical feedstocks. In this conference, we will present the optimisation model and the key findings from the case studies.

**References:**

[1] B.A. Excode (2016). Rice production and consumption trends in the Philippines from 2000 to 2015. Yuchengco Center, De La University; Philippines.

[2] C.T. Briones (Ed.) (2016). The PhilRice strategic plan 2017-2022. Philippine Rice Research Institute; Philippines.

[3] A.B. Mataia, A.B. Mataia, R.G. Manalili, J.C. Beltran, B.M. Catudan, N.M. Francisco and A.C. Flores (2016).

Analysis of the rice value chain in the Philippines. Philippine Rice Research Institute; Philippines

[4] R. Sims, A. Flammini, M. Puri, S. Bracco (2015). Opportunities for Agri-Food Chains to become energy-smart. Food and Agriculture Organization of the United Nations and the United States Agency for International Development; United States of America.

[5] S. Samsatli, N.J. Samsatli (2018). A multi-objective MILP model for the design and operation of future integrated multi-vector energy networks capturing detailed spatio-temporal dependencies. Applied Energy. DOI: 10.1016/j.apenergy.2017.09.055.

[6] S. Samsatli, N.J. Samsatli, N. Shah (2015). BVCM: a comprehensive and flexible toolkit for whole-system biomass value chain analysis and optimisation - mathematical formulation. Applied Energy, 147, pp. 131-160.

# 541194 Modeling the Impacts of International Food Trade on Contaminant Transport and Human Exposure


**Abstract:**

**Introduction:** Quantifying and addressing the implications of the transfer of resources across borders is a hallmark of research on the food, energy and water (FEW) nexus. International trade in goods can result in severe imbalances in the water stress, green house gas emissions, and energy consumption experienced by or attributed to producing and receiving countries[1,2]. Yet an important component of food trade—that is, the potentially hazardous chemicals contained in foods as a result of accumulation from production environments, processing, transport, or storage—is still poorly quantified[3–5]. Here, we investigate the importance of international food trade on the dietary exposure of the Swiss population to polybrominated diphenyl eithers (PBDEs). PBDEs are lipid soluble compounds used as flame retardants in synthetic fibers and molded plastics. Of major concern is PBDE exposure through the diet. Studies have confirmed that fish, meat, and dairy products contribute most to the daily PBDE intake[6]. Like other countries, Switzerland has reported a gradual shift in dietary patterns over the past few decades, with increasing consumption of fish reported, reaching a peak value of 17.6 kg in 2010, and with a shift in the origin of fish. We therefore constructed a model to quantify the exposure of the Swiss population to PBDEs from local and imported fish and seafood.

**Methodology:**

*Model boundaries:* Our study focused on Switzerland as a geographical case study. All calculations referred to fish imports for the year 2016. Fish, shrimp/prawn, and mussels were selected as key fish and seafood types to include in the study, as they are among the most highly consumed.

*Data collection:* An extensive literature review was conducted to derive useful input data for our model, including PBDE concentrations in fish, shrimp and mussels from different countries, and to estimate daily per capita fish consumption in Switzerland. Trade data (imports, exports, and re-exports) for fish, crustaceans (shrimp and prawns) and mollusks (mussels) was extracted from the UN Comtrade Database. Data were collected for fresh, frozen, fresh fillet, and frozen fillet categories of the respective commodities.

*Model Structure:* The model consists of four primary components. These are: (i) PBDE concentrations in fish from different origins, resulting from bioaccumulation from their environment and/or food; (ii) food trade data for Switzerland; (iii) local fish production; and (iv) dietary intake to the Swiss population via fish and seafood consumption.

**Results and Discussion:** Based on the difference between average per capita fish consumption, production, imports and re-exports, we estimate that Switzerland imports approximately 66% of fish consumed, while only 34% is locally caught. Of the imported fish, Norway, Denmark and the UK are the largest exporters of salmon to Switzerland. For shrimp and prawn, the majority is from Vietnam. For other fish, Norway, Denmark, the Netherlands, Vietnam, France, Germany, Italy, and China are the largest exporters. Within the locally produced fish, whitefish (*Coregonus sp.)* is the most common fish contributing to 60% of local the catch, followed by perch (19%), pike (7%) and roach (7%)[7]. Using an average of 23g/day fish consumption, results show that from 70000 metric tons fish consumed in Switzerland in 2016, 66 % (46000metric tons) was imported corresponding to 34 % locally caught fish (24000 metric tons).

Based on published studies of PBDEs in fish, the highest concentrations were detected in whitefish (Coregonus spp.) in Switzerland[8]. Norway also reported high levels of PBDEs in whitefish[9]. The US reported PBDEs as high as 3078 pg/g wet weight in salmon[10]. Compilation of country-specific PBDE body burdens in fish continues. In addition, we are employing modeling approaches to estimate PBDE body burdens in key species from exporting countries where data are lacking. These data will aid in forming a comprehensive source-receptor relationship for Swiss population exposure to PBDEs via seafood and fish.


**References Cited**


(1) Vora, N.; Shah, A.; Bilec, M. M.; Khanna, V. Food–Energy–Water Nexus: Quantifying Embodied Energy and GHG Emissions from Irrigation through Virtual Water Transfers in Food Trade. *ACS Sustain. Chem. Eng.* **2017**, *5* (3), 2119–2128.

(2) Leck, H.; Conway, D.; Bradshaw, M.; Rees, J. Tracing the Water–Energy–Food Nexus: Description, Theory and Practice. *Geogr. Compass* **2015**, *9* (8), 445–460.

(3) Ng, C. A.; von Goetz, N. The Global Food System as a Transport Pathway for Hazardous Chemicals: The Missing Link between Emissions and Exposure. *Environ. Health Perspect.* **2017**, *125* (1), 1–7.

(4) D'Odorico, P.; Carr, J. A.; Laio, F.; Ridolfi, L.; Vandoni, S. Feeding Humanity through Global Food Trade. *Earths Future* **2014**, *2* (9), 2014EF000250.

(5) Ercsey-Ravasz, M.; Toroczkai, Z.; Lakner, Z.; Baranyi, J. Complexity of the International Agro-Food Trade Network and Its Impact on Food Safety. *PLoS ONE* **2012**, *7* (5), 1–7.

(6) Schecter, A.; Haffner, D.; Colacino, J.; Patel, K.; Päpke, O.; Opel, M.; Birnbaum, L. Polybrominated Diphenyl Ethers (PBDEs) and Hexabromocyclodecane (HBCD) in Composite U.S. Food Samples. *Environ. Health Perspect.* **2010**, *118* (3), 357–362.

(7) Swiss fact: around 1.8 million kilograms of fish are caught in Switzerland every year https://lenews.ch/2017/01/04/swiss-fact-around-1-8-million-kilograms-of-fish-are-caught-in-switzerland-every-year/ (accessed May 31, 2018).

(8) Zennegg, M.; Kohler, M.; Gerecke, A. C.; Schmid, P. Polybrominated Diphenyl Ethers in Whitefish from Swiss Lakes and Farmed Rainbow Trout. *Chemosphere* **2003**, *51* (7), 545–553.

(9) Mariussen, E.; Fjeld, E.; Breivik, K.; Steinnes, E.; Borgen, A.; Kjellberg, G.; Schlabach, M. Elevated Levels of Polybrominated Diphenyl Ethers (PBDEs) in Fish from Lake Mjøsa, Norway. *Sci. Total Environ.* **2008**, *390* (1), 132–141.

(10) Schecter, A.; Päpke, O.; Tung, K.-C.; Staskal, D.; Birnbaum, L. Polybrominated Diphenyl Ethers Contamination of United States Food. *Environ. Sci. Technol.* **2004**, *38* (20), 5306–5311.

# 523949 Using Agricultural Wastes to Recover Rare Earth Elements from End-of-Life Materials

**Abstract:**

An economical and eco-friendly bench-scale process that relies on microbial production of organic acids has been developed to solubilize and release rare earth elements (REE) from solid recyclable materials (Reed, Hydrometallurgy, 2016, 166:34-40). Techno-economic analysis (TEA) and life cycle analysis (LCA) for bioleaching of REE from fluid catalytic cracking (FCC) catalysts suggested that a bioleaching plant could be profitable, but could be even more profitable if the cost of the sugar (glucose) used for growing the microbes was reduced (Thompson, ACS Sustainable Chemistry & Engineering, 2017, 6:1602-1609). This presentation will focus on our efforts to improve the economics of REE bioleaching using alternative carbon and energy sources as microbial growth substrates.

In our previous work examining the potential for several organic acid producing microorganisms to leach REE from end-of-life waste materials we identified *Gluconobacter oxydans* B58 as a promising industrial strain. Lixiviant produced by *G. oxydans,* containing predominantly gluconic acid, could recover ~50% of the total REE content of FCC catalyst. The biologically produced lixiviant was more effective than an abiotically prepared solution containing a similar concentration of gluconic acid. TEA and LCA showed that bioleaching could be more profitable and environmentally friendly than industrial standard chemical leaching processes, suggesting commercial viability. However, substantial reductions in operating costs could be achieved by replacing the refined glucose used in the original growth medium with a lower cost substrate. Consequently corn stover and potato wastewater were tested as growth substrates for *G. oxydans*. The initial lixiviants generated were tested against FCC and yielded a leaching efficiency 50-90% of the best case leaching efficiency of glucose-derived biolixiviant. Work is ongoing to optimize biolixiviant production from agricultural wastes. The conditions for improving bioleaching using waste-derived biolixiviants, including TEA and LCA evaluations, will be discussed.

# 526360 Membranes for Nutrient Concentration, Industrial Separations and Applications Beyond

**Abstract:**

With the increasing importance of the waterâ€“energyâ€“food nexus, it is critical that membranes can be designed for optimal performance and energy efficiency for large varieties of water and food applications. The cost and energy consumption of milk and whey concentration systems can be minimized when one selects a reverse osmosis element that reduces fluid flow requirements while maintaining optimum dewatering efficiency. In addition, superior performance of RO membranes which address the emerging importance of high rejection will be discussed for nutrient concentration, with a properly designed membrane based on the understanding of membrane structure-property relationships. Potentially there are many more applications where RO technology can be used to meet critical customer requirements with high efficiency and favorable economics.

# 543729 Dynamic Modulation of Protein Functions by Strand Displacement

**Abstract:**

Inspired by the remarkable ability of natural protein switches to sense and respond to a wide range of environmental queues, here we report a strategy to engineer synthetic protein switches by using a generalizable strand displacement format to dynamically organize proteins with highly diverse and complex architectures. We showed that both DNA and coiled-coil peptide strand displacements can be used to dynamically control the spatial proximity and the corresponding functions of different proteins for both in vitro and in vivo applications. The general utility of the strategy was used to performing Boolean logic operations to provide explicit control of protein function using multi-input, reversible, and amplification architectures suitable for a wide range of applications.

# 543730 Fabrication of Cellulosic Fibers for use as Functional Materials

**Abstract:**

Current growing demands in renewable materials have driven a new wave of technology development for processing of cellulosic fibers, targeting a broader range of applications far beyond the traditional paper and packaging material industries. In particular, surface affinity fabrication with engineered biomolecules and nanostructures have been showing promising in developing functional materials for selective absorbents, composites and electrochemical active materials. This talk will review recent advances in fabrication of cellulosic fibers for uses as functional materials.

# 543731 Nanomaterials for Combination Therapies and Immunomodulation

**Abstract:**

This talk will focus on how nanomaterials can be designed and developed to deliver combination therapies for cancer treatment, as well as activate the immune system and serve as adjuvants for sub-unit vaccines.

The first part of the talk will focus on how nanomaterials developed in our laboratory can be used to deliver combination therapies for the treatment of pancreatic cancer. Pancreatic cancer has a very high fatality rate, and the current treatment using gemcitabine (GEM) does not work effectively, partly due to desmoplasia. We have developed dual delivery nanoscale devices to deliver miRNA and GEM together, resulting in downregulation of the Sonic hedgehog signaling pathway and leading to inhibition of desmoplasia, pancreatic stellate cells and cancer stem cells. This, in turn, improved therapeutic outcomes of GEM in PC in mice through improving its perfusion in the tumor, and also led to significant reduction of metastasis.

The second part of the talk will focus on nanovaccines that can deliver pathogenic proteins, activate the immune system, and act as vaccine adjuvants. Respiratory pathogens such as influenza affect large segments of the population. Universal influenza vaccines that can provide effective protection against various strains of the pathogen, and in elderly populations as well, are sorely needed. We have developed virus-mimicking nanomaterials or nanovaccines that provide a depot for the antigen and enhance and modulate the immune response by targeting immune cells. We have shown long-term sustained neutralizing antibody titers, reduction of viral loads and effective protection against viral challenge in both young and aged mice as a result of administration of the nanovaccines.

# 543732 A Sense of Balance: Exploring the Role of Metabolic Pathway Modularization in the Microbial Production of Chemicals

**Abstract:**

Efforts in designing synthetic cells for metabolic engineering applications have been almost exclusively based on redirecting fluxes towards product synthesis by enzyme overexpression. However, such strategies result in draining large amounts of carbon precursors and energy resources (such as ATP and NAD(P)H) from the host. The imposed metabolic burden can cause unexpected physiological changes and dramatically reduces cell performance. For example, our simulations show that cell catabolism is abundant in wild type strains so that they can support a certain amount of metabolic burden without having apparent biosynthesis deficiency (e.g., without showing a slower growth after mutations). However, cell burden usually increases when genetic modifications are introduced or fermentative products are accumulated. When cell catabolism and respiration is unable to support the increasing ATP expenditure, the biosynthesis yield will suddenly drop, forming a "cliff" ("the straw that broke the camel's back").

In order to address the issue of metabolic burden, our laboratory has focused on developing novel approaches that allow metabolic pathway balancing and division of labor (DoL) in order to mitigate and if possible eliminate the negative effects of metabolic burden on cellular physiology. In one example, we have modularized the fatty acid biosynthetic pathway in E.coli, and improved the final titer as well as the production rate by balancing the different modules using different plasmid copy numbers and ribosome-binding sites of different strengths. A similar approach was used to balance a synthetic metabolic pathway that allows conversion of single-carbon (C1) substrates to multi-carbon natural products in a recombinant E.coli strain. In more recent research, we have used co-culture and poly-culture approaches where multistep pathways for producing phytochemicals have been split into different production strains in order to alleviate metabolic burden, allowing the de novo production of high-value chemicals from recombinant microorganisms for the first time. In this talk, I will highlight our recent work on engineering such strategies and its implication for industrial biomanufacturing.

# 543733 Repairing the Brain After Stroke: aÂ Biomaterials Strategy

**Abstract:**

Stroke is the leading cause of disability due to the brainâ€™s limited capacity to regenerate damaged tissue. After stroke, an increased inflammatory and immune response coupled with severely limited angiogenesis and neuronal growth results in a stroke cavity devoid of normal brain tissue. However, stroke also induces the formation of a pro-repair/plastic region in the area adjacent to the stroke core that commences in the weeks to days after stroke and subsides months after stroke. We propose that the delivery of pro-repair biomaterial scaffolds in synchronization with the endogenous pro-repair window will result in repair of the stroke cavity. This talk will focus on our current efforts to promote brain repair after stroke utilizing biomaterials.

# 537459 Bioprocess Platform for High Cell Density Cultivation for Probiotic Yeast Production in Semi-Industrial Scale

**Abstract:**

Based on the latest information about human microbiome, the number of mammalian cells contributes to only 42% of the total of healthy human body. The rest of cells are a consortium of microbial cells of different groups who play major role in human healthy life. These microbes are commonly named as probiotics and contribute in the production of many bioactive metabolites in human body with many functional roles. They act as source of vitamins, immune system enhancers, compete with microbial pathogens through antibiotic production, and many other bioactivities. Therefore, continuous supplementation of human body with these beneficial microbes is necessary to sustain the healthy life of individual. For many years, industrial production of bacterial probiotics belonging to *Lactobacillus, Bifidobacterium* and *Bacillus* was extensively studied. However, little information is available for large scale production of probiotic yeasts. In this work, techno-industrial platform design for production of three types of probiotic yeasts will be presented namely: *Saccharomyces boulardii* and *Klyveromyces lactis*, and selenium enriched probiotic *Saccharomyces boulardii.* Different methods of bioprocessing optimizations and cultivation strategies for each type of cells have been studied to improve biomass production in small scale. Furthermore, the processes have been scaled up to semi-industrial scale production to develop complete industrial platform.

# 548064 Metabolic Engineering of *Clostridium cellulovorus* for *n*-Butanol Production from Cellulosic Biomass in a Consolidated Bioprocess

**Abstract:**

*n*-Butanol can be used as a significant industrial chemical and potential superior biofuel. Especially, compared with ethanol, *n*-butanol can be used as an ideal substitute of gasoline because of its high energy density, low water solubility, and low vapor pressure. Lignocellulosic biomass is an abundant, cheap, and renewable carbon source, and is a desirable feedstock for biofuels production. In our previous research, *Clostridium cellulovorans*, a natural cellulose/hemicelluloses utilizing, butyrate producing bacterium, was metabolically engineered for n-butanol production from cellulose by overexpressing *adhE2* from *Clostridium acetobutylicum*. Therefore, consolidated bioprocessing (CBP) with combined enzyme production, biomass hydrolysis, and sugar fermentation in one step can be achieved by using the engineered *C. cellulovorans*. However, this mutant only produced 1.42 g/L n-butanol and 1.60 g/L ethanol from microcrystalline cellulose in 10 days. In this study, genome-scale metabolic model with omics analysis was firstly applied for guiding rational metabolic engineering of *C. cellulovorans*. System metabolic and process engineering, including optimization of butanol biosynthesis pathway, increasing C4 carbon flux, weakening competitive acids synthesis pathway, and improving the intracellular NADH availability through redox rebalancing, were carried out to increase butanol production in *C. cellulovorans*. The mutant produced 5.2 g/L *n*-butanol with a high yield of >0.3 g/g cellulose in batch fermentation, which were the highest ever achieved. This study demonstrated that *C. cellulovorans* is a promising CBP host for biobutanol production from cellulosic biomass.

# 523681 Phenotypic Adaptation of a Novel Bacterium for a Low-Cost Production of D-Lactic Acid

**Abstract:**

In bioplastic industry, we can divide the processing technology into 3 catergories including upstream, intermediate, and downstream processes. In the upstream process, we can subcategorize into 3 main steps, i.e., feedstocks pretreatment, fermentation, and downstream product recovery and purification. The performance of the upstream process can be justified by low production cost, consumption of renewable, nonedible feedstocks that are locally available. Currently, the edible feedstocks are still used in the conventional fermentation process. This is not only affecting the customers in terms of the demand/supply value chain, but it is also directly contributing on the high sale price of the food products as well as the monomers for bioplastic synthesis. To minimize the problems earlier stated, using the renewable, nonedible feedstocks in the production of monomers for bioplastic is promising. In addition, the high process performance is required. Key factors to achieve high production performance include robust microorganisms and simple but effective processes. The robust microbes can be found in nature or they can be metabolically engineered to the desired purposes. *Terrilactibacillus laevilacticus* SK5-6, a novel D-lactate producing bacterium was used in this work. The wild type *T. laevilacticus* SK5-6 was adapted in the preculture medium with the low concentration of yeast extract. From the experimental results, *T. laevilactibacillus* SK5-6 was adapted in the medium containing the very low yeast extract concentration of 0.45 g/L (97% reduction from the base case medium). The viable fermentation platform was developed in the 5 L stirred fermentor. The seed train was successfully developed for further fermentation scale up in the 30 L stirred fermentor. The obtained fermentation kinetics from the stirred fermentors (both 5 L and 30 L) were comparable to the theoretical values. The simplified economic evaluation was conducted. It was found that the raw material costs in D-lactic acid fermentation by the wild type strain (SK5-6) and the adapted strain (SK5-6T20) were low in comparable to the sale price of D-lactic acid in the market. From the findings previously mentioned, this indicates that the developing platform in this study is promising.

# 527300 Development of a Genome-Scale Metabolic Model for *S. Cerevisiae* to Facilitate Understanding of theÂ Differences in Metabolism between Commercial Yeast Strains

**Abstract:**

Two key metabolic activities relevant to industrial wine fermentations are nutrient utilization efficiency and tolerance to high ethanol concentrations exhibited by industrial yeast strains. Therefore, to study the details of yeast metabolism, is of great interest to develop ways to control stuck or sluggish fermentations. One approach is to use computational methods due to their advantage of being comprehensive and more economical compared to experimental methods. Hence, many studies have been conducted to create genome-scale metabolic models of yeast. Despite progress in the field, most current models either focus on aerobic systems or lack the detailed lipid metabolism that has been shown experimentally to be highly correlated with nutrient utilization efficiency. One way to capture the power of these models is to use (dynamic) FBA (flux balance analysis) to predict the flux distribution of all the metabolites within the cell over the course of an entire fermentation. Using this approach, it is possible to test the predictive capability of these models by comparing predictions with experimental fermentation data. Once the models fit dynamic data, they can be used to understand differences between commercial strains and suggest genetic modification strategies towards increasing strain ethanol tolerance and nutrient utilization efficiency. In this study, we improve the latest consensus genome scale model of yeast by incorporating additional lipid pathways. Previously, we showed that nutrient utilization efficiency and ethanol tolerance of 22 different industrial yeast strains were a strong function their lipid composition while molecular mechanisms of these phenomena were not elucidated. By utilizing the latest consensus genome-scale metabolic reconstruction of yeast, which has the most comprehensive representation of fatty acid, glycerolipid, and glycerophospholipid metabolism, we can more accurately predict metabolic fluxes for various yeast strains. In turn, understand the variation in metabolism between different strains that lead to disparities in nutrient utilization efficiency and aroma production.

# 527524 Kinetics of Cell Growth and Invertase Production By the Biotherapeutic Yeast, *Saccharomyces Boulardii*

**Abstract:**

Invertase (Î²-D-fructofuranoside fructohydrolase; EC 3.2.1.26) is a highly glycosylated enzyme which hydrolyzes sucrose to glucose and fructose. Therefore, it finds many applications in food and feed industries. In addition, this enzyme become more attractive for its potential application in pharmaceutical industries based on its antimicrobial activities and its ability to reduce the side effects of cancer therapy. In this work, the biotherapeutic yeast *Saccharomyces boulardii* has been used as a potential biofactory for extracellular invertase production using semi-defind cultivation medium in different submerged cultivation systems. The initial results showed that the optimal sucrose concentration for invertase production was 30 g $L^{-1}$ which yielded volumetric enzyme production of 6540 U $L^{-1}$. Further studies for bioprocess optimization were carried out in shake flask, bioreactor cultivations under uncontrolled and controlled pH conditions. The results clearly demonstrated that, scaling up of process from shake flask to bioreactor level increased volumetric enzyme production up to 8111 U $L^{-1}$. However, this increase was due to the increase in biomass production (from 3.6 up to 4.4 g $L^{-1}$) rather than cell productivity as both cultures showed almost the same specific enzyme production of about 1980 U $g^{-1}$. Further improvement in the production process was achieved in pH-controlled bioreactor culture (pH 5.5). The maximal invertase production in the controlled culture was increased up to 14830 U $L^{-1}$ concomitant with a significant increase in biomass up to 7.5 g $L^{-1}$. However, on calculating the specific growth rate, the specific enzyme production was about 2269 U $g^{-1}$. Thus, we can conclude that the increase in invertase production in bioreactor under controlled pH condition was not only due to the increase in biomass but also due to the increase of cell productivity. The process was further developed through cultivation under different aeration rates. The results showed that aeration at 1 v $v^{-1}$ $m^{-1}$ was the most suitable for invertase production in batch as well as complete medium-fed cultivations.

# 527848 L-Lactate Production By a Potent Homofermentative *Bacillus* Sp. BC-001

**Abstract:**

A novel *Bacillus* sp. BC-001 was found as a potent thermotolerant and homofermentative L-lactic acid production. This strain wasisolated from the natural resources in Thailand. From the screening fermentation in the flask culture, this novel strain produced L-lactate with high yield and productivity during batch cultivation of glucose (0.96 g/g and 0.87 g/L·h, respectively). Not only glucose, BC-001 could utilize several carbon substrates for lactic acid including arabinose, xylose, sucrose, and fructose. From the preliminary data obtained in first screening test, we proposed this study to optimize the process conditions to achieve a high production performance of L-lactate fermentation by this novel strain. The results in the flask culture revealed that a proper seed culture was obtained at the correct physiological stage with adequate mixing and sufficiently high air permeability. This resulted in the high production of lactic acid. Further optimization was conducted in a 5-L stirred fermentor employing the high/heavy-inoculum concepts. It was found that the high-inoculum (71% inoculum size) was significant increase in lactate productivity (5.5-6.1 g/L·h). Interestingly, BC-001 could use both calcium bases and monovalent based for pH control during lactate fermentation in the 5-L stirred fermentor. This indicated that the fermentation process was easy to control and operate. To confirm BC-001 performance as the industrial L-lactate producer, the fermentation platform was conducted in 3 series of the 30, 300, and 3,000 L stirred fermentor. To control the phenomena of lactate production performance in scale-up fermentor, the power input per unit volume was selected as the constant criterion. From the pilot scale test, we obtained the fermentation reproducibility in the 3 series of the fermentation runs.

# 532420 (Keynote) Towards Efficient Bioproduction of Polymalic Acid and Malic Acid Production: System Metabolic and Process Engineering

**Abstract:**

Polymalic acid (PMA) is a novel polyester polymer that has been broadly used in the medical and food industries. Its monomer, L-malic acid (MA), is also a potential C4 platform chemical. Recently, great researches have been attempted to develop biotechnological process for PMA and L-malic acid production. Here, we integrated multi-omics, genome-scale metabolic model and genome-editing technique to engineer the metabolic nodes for efficient PMA production. Reductive TCA cycle and glyoxylate shunt were verified that could strengthen carbon flux for PMA and MA synthesis. Subsequently, some nodes, including pyruvate carboxylase gene (*pyc*), malate synthase gene (*mls*), isocitrate lyase gene (*icl*), and carbon response transcription factor *cat8* gene, were overexpressed, respectively. Among of them, the OE::*mls-icl-cat8* strain was showed the greatest PMA titer and yield (37.94g/L and 0.45g/g glucose) in shake-flask, and increased by 26% and 25% compared with the wild type strain, respectively. CRISPR/Cas9 system was successfully developed and applied as a simple and powerful tool for targeted gene knock-out and/or knock-in *A. pullulans*. Some renewable biomass feedstocks, i.e. sugarcane molasses and xylose mother liquor, were also presented.

# AUTHOR INDEX

# AUTHOR INDEX

# AUTHOR INDEX

# AUTHOR INDEX